# Advances in Intelligent Systems and Computing

Volume 897

**Series editor**

Janusz Kacprzyk, Systems Research Institute, Polish Academy of Sciences, Warsaw, Poland
e-mail: kacprzyk@ibspan.waw.pl

The series "Advances in Intelligent Systems and Computing" contains publications on theory, applications, and design methods of Intelligent Systems and Intelligent Computing. Virtually all disciplines such as engineering, natural sciences, computer and information science, ICT, economics, business, e-commerce, environment, healthcare, life science are covered. The list of topics spans all the areas of modern intelligent systems and computing such as: computational intelligence, soft computing including neural networks, fuzzy systems, evolutionary computing and the fusion of these paradigms, social intelligence, ambient intelligence, computational neuroscience, artificial life, virtual worlds and society, cognitive science and systems, Perception and Vision, DNA and immune based systems, self-organizing and adaptive systems, e-Learning and teaching, human-centered and human-centric computing, recommender systems, intelligent control, robotics and mechatronics including human-machine teaming, knowledge-based paradigms, learning paradigms, machine ethics, intelligent data analysis, knowledge management, intelligent agents, intelligent decision making and support, intelligent network security, trust management, interactive entertainment, Web intelligence and multimedia.

The publications within "Advances in Intelligent Systems and Computing" are primarily proceedings of important conferences, symposia and congresses. They cover significant recent developments in the field, both of a foundational and applicable character. An important characteristic feature of the series is the short publication time and world-wide distribution. This permits a rapid and broad dissemination of research results.



More information about this series at http://www.springer.com/series/11156

Rituparna Chaki · Agostino Cortesi
Khalid Saeed · Nabendu Chaki
Editors

# Advanced Computing and Systems for Security

## Volume Seven

*Editors*
Rituparna Chaki
A.K. Choudhury School of Information Technology
University of Calcutta
Kolkata, West Bengal, India

Agostino Cortesi
Dipartimento di Scienze Ambientali, Informatica e Statistica
Università Ca' Foscari Venezia
Mestre, Venice, Italy

Khalid Saeed
Faculty of Computer Science
Bialystok University of Technology
Bialystok, Poland

Nabendu Chaki
Department of Computer Science and Engineering
University of Calcutta
Kolkata, West Bengal, India

ISSN 2194-5357 ISSN 2194-5365 (electronic)
Advances in Intelligent Systems and Computing
ISBN 978-981-13-3249-4 ISBN 978-981-13-3250-0 (eBook)
https://doi.org/10.1007/978-981-13-3250-0

Library of Congress Control Number: 2018961203



This Springer imprint is published by the registered company Springer Nature Singapore Pte Ltd.
The registered company address is: 152 Beach Road, #21-01/04 Gateway East, Singapore 189721, Singapore

# Preface

The fifth International Doctoral Symposium on Applied Computation and Security Systems (ACSS 2018) was held in Kolkata, India, during February 9–11, 2018. The University of Calcutta in collaboration with Ca Foscari University of Venice, Italy, and Bialystok University of Technology, Poland, organized the symposium.

The fifth year of the symposium was marked with a significant shift of research interests toward usage of machine learning for signal processing and image analytics. It is truly interesting to find the increasing popularity of ACSS among researchers, this being one of its own kinds of the symposium for doctoral students to showcase their ongoing research works on a global platform. The program committee members are each renowned researcher in his/her field of interest, and we thank them for taking immense care in finding out the pros and cons of each of the submissions for ACSS 2018. As in previous years, the session chairs for each session had a prior go-through of each paper to be presented during the respective sessions. This often makes it more interesting as we found deep involvement of the session chairs in mentoring the young scholars during their presentations. With concrete suggestion on how to improve the presented works, the participants utilized the 6-week post-symposium time. The final version of the papers thus goes through the second level of modification as per the session chair's comments.

These post-symposium book volumes contain the revised and improved version of the manuscripts of works presented during the symposium. The evolution of ACSS is an ongoing process. We had included deep learning in the scope of research interests in 2018. In 2019, considering the global interest, we are planning to include pricing and network economics within the scope and announced the same in CFP for the next year. We have invited researchers working in the domains of algorithms, signal processing and analytics, security, image processing, and IoT to submit their ongoing research works.

The indexing initiatives from Springer have drawn a large number of high-quality submissions from scholars in India and abroad. We are indebted to all the program committee members, who, despite their immensely busy schedules, have given an intense reading of each allotted contribution. Each reviewer has given their constructive suggestions for the submitted works. ACSS continues with the tradition of the double-blind review process by the PC members and by external reviewers. The reviewers mainly considered the technical aspect and novelty of each paper, besides the validation of each work. This being a doctoral symposium, the clarity of presentation was also given importance. The entire process of paper submission, review, and acceptance process was done electronically. Due to the sincere efforts of the program committee and of the organizing committee members, the symposium resulted in a suite of strong technical paper presentations followed by effective discussions and suggestions for improvement for each researcher.

The Technical Program Committee of the symposium selected only 24 papers for publication out of 64 submissions. We would like to take this opportunity to thank all the members of the program committee and the external reviewers for their excellent and time-bound review works. We thank all the sponsors who have come forward toward the organization of this symposium. These include Tata Consultancy Services (TCS), Springer Nature, ACM India, M/s Fujitsu, Inc., India. We appreciate the initiative and support from Mr. Aninda Bose and his colleagues in Springer Nature for their strong support toward publishing this post-symposium book in the series "Advances in Intelligent Systems and Computing.". Last, but not least, we thank all the authors without whom the symposium would not have reached up to this standard.

On behalf of the editorial team of ACSS 2018, we sincerely hope that ACSS 2018 and the works discussed in the symposium will be beneficial to all its readers and motivate them toward even better works.

Kolkata, West Bengal, India — Rituparna Chaki
Bialystok, Poland — Khalid Saeed
Mestre, Venice, Italy — Agostino Cortesi
Kolkata, West Bengal, India — Nabendu Chaki

# Contents


**Part I Algorithms**

**Gossip-Based Real-Time Task Scheduling Using Expander Graph** . . . . . 3
Moumita Chatterjee and S. K. Setua

**Exact Algorithm for L(2, 1) Labeling of Cartesian Product Between Complete Bipartite Graph and Path** . . . . . 15
Sumonta Ghosh and Anita Pal

**Extraction and Classification of Blood Vessel Minutiae in the Image of a Diseased Human Retina** . . . . . 27
Piotr Szymkowski and Khalid Saeed

**Part II Signal Processing and Analytics—I**

**Analysis of Stimuli Discrimination in Indian Patients with Chronic Schizophrenia** . . . . . 49
Jaskirat Singh, Sukhwinder Singh, Savita Gupta and Bir Singh Chavan

**Analysis of Resting State EEG Signals of Adults with Attention-Deficit Hyperactivity Disorder** . . . . . 61
Simranjit Kaur, Sukhwinder Singh, Priti Arun and Damanjeet Kaur

**Readability Analysis of Textual Content Using Eye Tracking** . . . . . 73
Aniruddha Sinha, Rikayan Chaki, Bikram De Kumar and Sanjoy Kumar Saha

**Part III Signal Processing and Analytics—II**

**Multi-node Approach for Map Data Processing** . . . . . 91
Vít Ptošek and Kateřina Slaninová

**Naive Bayes and Decision Tree Classifier for Streaming Data Using HBase** . . . . . 105
Aradhita Mukherjee, Sudip Mondal, Nabendu Chaki and Sunirmal Khatua

**FPGA-Based Novel Speech Enhancement System Using Microphone Activity Detector** . . . . . 117
Tanmay Biswas, Shuvadeep Bhattacharjee, Sudhindu Bikash Mandal, Debasri Saha and Amlan Chakrabarti

**Part IV Software and Service Engineering**

**Optimal Mapping of Applications on Data Centers in Sensor-Cloud Environment** . . . . . 131
Biplab Kanti Sen, Sunirmal Khatua and Rajib K. Das

**Constraint Specification for Service-Oriented Architecture** . . . . . 143
Shreya Banerjee, Shruti Bajpai and Anirban Sarkar

**Software Regression and Migration Assistance Using Dynamic Instrumentation** . . . . . 159
Nachiketa Chatterjee, Amlan Chakrabarti and Partha Pratim Das

**Author Index** . . . . . 171

# About the Editors

**Rituparna Chaki** is Professor of Information Technology at the University of Calcutta, India. She received her Ph.D. from Jadavpur University in India in System Executive in the Ministry of Steel, Government of India, for 9 years, before joining academia in 2005 as Reader of Computer Science and Engineering at the West Bengal University of Technology, India. She has been with the University of Calcutta since 2013. Her areas of research include optical networks, sensor networks, mobile ad hoc networks, Internet of things, and data mining. She has nearly 100 publications to her credit. She has also served on the program committees of various international conferences and has been Regular Visiting Professor at the AGH University of Science and Technology, Poland, for last few years. She has co-authored a couple of books published by CRC Press, USA.

**Agostino Cortesi, Ph.D.,** is Full Professor of Computer Science at Ca' Foscari University, Venice, Italy. He served as Dean of Computer Science Studies, as Department Chair, and as Vice-Rector for quality assessment and institutional affairs. His research interests concern programming languages theory, software engineering, and static analysis techniques, with a focus on security applications. He has published more than 110 papers in high-level international journals and international conference proceedings. His h-index is 16 according to Scopus and 24 according to Google Scholar. He served several times as a member (or chair) of program committees of international conferences (e.g., SAS, VMCAI, CSF, CISIM, ACM SAC), and he is on the editorial boards of the journals *Computer Languages, Systems and Structures* and *Journal of Universal Computer Science*. Currently, he holds the chairs of "Software Engineering," "Program Analysis and Verification," "Computer Networks and Information Systems," and "Data Programming."

**Khalid Saeed** is Full Professor in the Faculty of Computer Science, Bialystok University of Technology, Bialystok, Poland. He received his B.Sc. in electrical and electronics engineering from Baghdad University in 1976 and M.Sc. and Ph.D. from Wroclaw University of Technology in Poland in 1978 and 1981, respectively.

He received his D.Sc. in computer science from the Polish Academy of Sciences in Warsaw in 2007. He was Visiting Professor of Computer Science at the Bialystok University of Technology, where he is now working as Full Professor. He was with the AGH University of Science and Technology from 2008 to 2014. He is also working as Professor in the Faculty of Mathematics and Information Sciences at Warsaw University of Technology. His areas of interest are biometrics, image analysis and processing, and computer information systems. He has published more than 220 papers in journals and conference proceedings and edited 28 books, including 10 textbooks and reference books. He supervised more than 130 M.Sc. and 16 Ph.D. theses and given more than 40 invited lectures and keynotes in Europe, China, India, South Korea, and Japan. He has received more than 20 academic awards and is a member of the editorial boards of over 20 international journals and conferences. He is IEEE Senior Member and was selected as IEEE Distinguished Speaker for 2011–2016. He is Editor-in-Chief of International Journal of Biometrics, published by Inderscience.

**Nabendu Chaki** is Professor in the Department of Computer Science and Engineering, University of Calcutta, Kolkata, India. He first graduated in physics from the legendary Presidency College in Kolkata and then in computer science and engineering from the University of Calcutta. He completed his Ph.D. from Jadavpur University, India, in 2000. He shares six international patents, including four US patents, with his students. He has been active in developing international standards for software engineering and cloud computing as Member of the Global Directory (GD) for ISO-IEC. As well as editing more than 25 book volumes, he has authored 6 textbooks and research books and has published over 150 Scopus-indexed research papers in journals and at international conferences. His areas of research interest include distributed systems, image processing, and software engineering. He also served as Researcher in the Ph.D. program in software engineering at the U.S. Naval Postgraduate School, Monterey, CA. He is a visiting faculty member for numerous universities in India and abroad. In addition to serving on the editorial board for several international journals, he has also been on the committees of over 50 international conferences. He is Founder Chair of ACM Professional Chapter in Kolkata.

# Part I
# Algorithms

# Gossip-Based Real-Time Task Scheduling Using Expander Graph

**Moumita Chatterjee and S. K. Setua**

**Abstract** In this paper, we consider the scheduling of real-time distributed tasks in large-scale dynamic networks, where node and link failures and message losses occur frequently. We propose a distributed scheduling algorithm using gossip-based approach called GBTS for dynamic and reliable discovery of suitable nodes that can execute the tasks. GBTS takes advantage of the slack times to optimize the gossiping duration, thereby satisfying the end-to-end timing constraints of tasks with a probabilistic guarantee. Although gossip-based protocols are fault tolerant, they incur high message overheads. We propose to use a highly connected, sparse graph called the expander graph to control the communication complexity of our algorithm. Performance analysis shows that GBTS performs better in terms of both time and message complexity.

**Keywords** Time/utility functions · Real-time task · Gossip
Transition probability · Expander graphs

## 1 Introduction

A real-time task is characterized by its arrival time, execution time, and deadline. Researches on real-time task scheduling have proposed algorithms that ensure or optimize compliance with deadlines. In a distributed system, a task is created by a node as a response to some events. After creation, each task requires a set of data objects, distributed over various nodes in the network to complete its execution. For each task, the current executing node also called the head node tries to locate the next node in the network at which the next section of the task will be executed. In a large-scale system, the set of processors which are available for computation may

M. Chatterjee (✉) · S. K. Setua
Department of Computer Science and Engineering, University of Calcutta, Kolkata, India
e-mail: moumitachatterji@gmail.com

S. K. Setua
e-mail: sksetua@gmail.com



R. Chaki et al. (eds.), *Advanced Computing and Systems for Security*,
Advances in Intelligent Systems and Computing 897,
https://doi.org/10.1007/978-981-13-3250-0_1

become unavailable due to failures. In these systems, the nodes containing the data objects may leave the network or may fail. So each head node has to dynamically determine the next node on which the next requested object is located and send the task section for execution within its stipulated deadline.

A lot of research works [1, 2] etc., have focused on the problem of real-time end-to-end scheduling and timing guarantees in unreliable networks. But they have not dealt with the message overhead problem, which is the main goal of our work. RTG-L [3] guarantees that real-time threads can meet their deadlines in a large-scale dynamic network. RTG-L communicates among the nodes using gossip-based protocol. Gossip algorithms are robust against node and link failures, scalable, and fault tolerant, and do not require any error recovery mechanism. Although the message propagation using gossip is usually probabilistic in nature, the protocol achieves high stability under network disruptions and scales to a large number of nodes. However, gossip-based algorithms incur high message overheads. RTQG [4] which builds on RTG-L reduces the number of messages used for communication by using quorum system for communication.

In this paper, we propose a gossip-based task scheduling algorithm (GBTS) to schedule real-time tasks in a large-scale dynamic networks. Although GBTS is also a gossip-based algorithm, it is different from the previous works [3, 4] as it uses a different message propagation approach for achieving better communication complexity as compared to the previous protocols. We propose to use as communication graph a highly connected sparse graph known as the expander graph. Expander graphs can guarantee connectivity of the network and reliable communication in presence of arbitrary (Byzantine) failures of nodes in the network. Another feature of our algorithm is that GBTS uses a better performance metric compared to the previous protocols for scheduling real-time tasks locally at a node. GBTS schedules tasks in such a way, so that the slack time of the tasks can be utilized for gossiping.

The rest of the paper is organized as follows: Sect. 2 describes the models and objectives of work. Our proposed approach is described in Sect. 3. The performance analysis of our algorithm is described in Sect. 4. Section 5 concludes the paper.

## 2 Models and Objectives of Work

### *2.1 Real-Time Distributable Tasks*

A task is the basic unit of execution in a distributed system. A task is created by a node at arbitrary times as a response to some events. Each task references a set of data objects distributed over various nodes in the network to complete its execution. We define a task $T_i$ as a multi-tuple as follows: $<TID_i, <t_{ik}, O_j, ER_{ik}, size_{ik}>, T_i^{EX}, NS_i, DL_i>$ where

- $TID_i$—is the identifier of the task $T_i$.

- <$t_{ik}$, $O_j$, $ER_{ik}$, $size_{ik}$> denotes that the section of the task $t_{ik}$ requires object $O_j$ for its execution, the execution time for this section of the task at node j is $ER_{ik}$ and size of the task is $size_{ik}$.
- $TEX_i$—Total Execution Time of the task $T_i$.
- $NS_i$—denotes the total number of task sections $T_i$.
- $DL_i$—denotes the deadline of the task $T_i$.

The portion of the task $t_{ik}$ executing object $O_j$ is known as the *k*th section of the task. Thus, we can view a task as a collection of task sections. The initial section of a task is called its root and the recent active section of the task is called its head. The identifier of the nodes holding the objects is not known in advance as nodes may dynamically join or leave the network or fail due to some reasons.

## 2.2 Timeliness Model and Utility Enhancement Scheduling

The time constraint of a task is specified using time/utility function (TUF) [5]. TUF is used to normalize the deadline constraint of real-time tasks. A TUF represents the utility to the system resulting from the completion of the task as a function of its completion time. A TUF signifies the urgency of a task measured as the deadline on the X-axis and its importance measured as a utility function on the Y-axis. A task $T_i$s TUF is denoted as $U_i(t)$. A classical deadline is unit-valued, i.e., $U_i(t) = \{0, 1\}$ and a downward step TUF as the utility is not considered. The classical deadline is generalized by the downward step TUF, where $U_i(t) = \{0, n\}$. Sections of the same task have the same TUFs. The TUF of the *j*th section of $T_i$ is denoted by $U_{ij}(t)$. In this paper, we restrict our focus on downward step TUFs.

Each TUF $U_i$, $i \in \{1, \ldots, n\}$ has an initial time $Ar_i$ and a deadline $Dl_i$. The initial time is the earliest time for which the function is defined while the deadline is the latest time at which a downward step TUF reaches zero utility value. That is, we can write that for all tasks i, $U_i(t) > 0$ for all t in the range of $[Ar_i, Dl_i]$ and $U_i(t) = 0$ otherwise. If the deadline is reached and a task has not completed its execution, then an exception is issued. This exception causes the task to be aborted and issues exception handlers to abort the partially executed sections of the task (this is done to release system resources). The time constraint of the exception handlers is also specified using TUFs.

## 2.3 System Model

We consider a distributed network of n nodes denoted by $V(t) = \{1, \ldots, n\}$. The nodes in the network can communicate with each other through bidirectional links denoted by E(t). We consider a failure prone network, where each node can fail and can recover independently and the communication links between the nodes may also fail while delivering messages. We also assume network partitioning, where nodes and link may fail concurrently and nodes in different partitions may become inaccessible to each other. We can define such a dynamic network as follows:

$$DyNet(t) = < G(t) = \big(V(t),\ E(t),\ f_i(v,t),\ e_j(e,t)\big) > \tag{1}$$

$f_i$: $V(t) = Val_i$—denotes the $i$th function on the graphs vertices, where $i \in \{1, \ldots, VN\}$. VN is the number of vertex functions.

$e_j$: $E(t) = Val_j$ denotes the $j$th function on the graphs edges, where $j \in \{1, \ldots, EN\}$. EN is the number of edge functions.

The functions $f_i$ and $e_j$ are assumed to change from time to time. We can represent such a graph by an adjacency matrix A(t) which is a symmetric matrix of the form $V(t) \times V(t)$ where each element $A_{ij}$ is such that

$$A_{ij} = \begin{cases} 1 \text{ if there exists an edge between vertices i and j at time t} \\ 0 \text{ otherwise} \end{cases} \tag{2}$$

The set of vertices to which a node $v_i$ is directly connected at time t is defined as the first neighbors of $v_i$, [6] i.e.,

$$\Gamma_i^1(t) = \{v_j \in V(t) : \{v_i, v_j\} \in E(t)\} \tag{3}$$

The degree of node $v_i$ at time t is defined as the number of nodes in the first neighborhood of $v_i$, i.e.,

$$\beta_i(t) = \left|\Gamma_i^1(t)\right| \tag{4}$$

Let $\delta = min_{i \in V(t)} \beta_i(t)$ be the minimum degree, $d = 2|E(t)|/n$ be the average degree of the network and $\Delta = max_{i \in V(t)} \beta_i(t)$ be the maximum degree.

A path between vertex $v_i$ and $v_j$ is a sequence of vertices $< v_0, v_1, \ldots, v_{k-1}, v_k >$, where $v_{i-1}, v_i \in E(t)$ for all i = 1, …, k. The number of edges in the path is known as the length of the path. The shortest path between any two vertices $v_i$ and $v_j$ is known as $dist_{ij}(t)$. We use $\Gamma_i^k(t)$ to denote the set of nodes at a distance of k from node $v_{i,}$ i.e.,

$$\Gamma_i^k(t) = \{v_j \in V(t) | dist_{ij}(t) = k\} \tag{5}$$

The diameter D(t) is defined as the longest shortest path.

$$D(t) = max_{vi,vj \in V(t)} \{dist_{ij}(t)\} \tag{6}$$

Let $N_i(t)$ denote the number of actual edges that the first neighborhood of vertex $v_i$ has, i.e., $N_i(t)$ is given by Kasprzyk [6]

$$N_i(t) = |\{v_l, v_k\} : v_l, v_k \in \Gamma_i^1(t) \wedge \{v_l, v_k\} \in E(t)\}| \quad (7)$$

## 2.4 *Objectives and Contribution of the Work*

Our objective in this work is to design an algorithm for scheduling tasks in a network with node/link failures and message losses and demonstrate the effectiveness of the new approach. The major contributions of the paper are as follows:

1. We propose a new algorithm using gossip for scheduling real-time tasks in a network with node and link failures.
2. We use the time utility function to define a new metric to locally schedule each task section at a node, so as to increase the total utility for each section and reducing the number of abortions.
3. We demonstrate that by using expander graphs to communicate among the nodes requires minimum time and communication complexity compared to other approaches.

## 2.5 *Real-Time Task Scheduling Using Gossip*

When a node issues a task for execution, the task runs on that node (also known as the present head node of the task) until the task requires a different set of objects for its execution. In this case, the current head node of the task must determine the next head node of the task, i.e., the node on which the next section of the task will run. As nodes in the network may fail or may leave the network at any time during the computation, the present head node of the task must dynamically determine the next head node on which the next section of the task can execute. Thus, head nodes use gossip protocols to reliably spread information over the network, so that head nodes can determine the next successor of the task despite node failures and losses.

# 3 Gossip-Based Task Scheduling Algorithm (GBTS)

The algorithm consists of two phases at which it is described: local phase and global phase.

## 3.1 Local Phase

When a task section $t_{ik}$ of a task $T_i$ arrives at node j at time t the following values are calculated:

**Step 1: Time utility function calculation for each task section $t_{ik}$**

i. Number of sections of the task $NS_i$ is reduced by 1, i.e., $NS_i = NS_i - 1$
ii. Total remaining slack time $SLACK_i^{Total}$ of task $T_i$

$$SLACK_i^{Total} = DL_i - t - (ER_{ik} - t) \tag{8}$$

iii. Calculate local slack time $LS_{ij}$ for task $T_i$

$$LS_{ij} = \begin{cases} \frac{STACK_i^{Total}}{NS_i} & \text{if } NS_i > 1 \\ STACK_i^{Total} & \text{if } 0 \le NS_i \le 1 \end{cases} \tag{9}$$

iv. Calculate local termination time of task section $t_{ik}$

$$TER_k^i = t + ER_{ik} + LS_{ij} \tag{10}$$

v. GBTS then divides the remaining slack time equally among the remaining task sections to give them a fair chance to complete their execution. This local termination time is used by the algorithm to test for schedule feasibility while constructing local schedules.

**Step 2: Estimating the resource consumption of each task section $t_{ik}$**

The memory consumed by the task section $t_{ik}$ is given by the $size_{ik}$. The CPU consumption is given by its execution time $ER_{ik}$. The consumption of a node's resource by a task section $t_{ik}$ is given as a product of its CPU and memory utilization. High value for this metric denotes a high CPU and memory consumption. The CPU and memory consumption is calculated using the following formula:

$$CONS_{memory}^{CPU} = ER_{ik} \times size_{ik} \tag{11}$$

**Step 3: Utility and consumption ratio calculation**

We define a performance metric utility and consumption ratio (UCR) to incorporate the timeliness and the CPU and memory consumption of a task section. The UCR of a task section measures the utility that can be gained by executing that task section per unit CPU and memory consumption. The UCR of a task section $t_{ik}$ at time t is given by

$$UCR_{it}(t) = \frac{(t + TER_k^i)}{CONS_{memory}^{CPU}} \tag{12}$$

**Step 4: Constructing local schedules**

In this step, GBTS constructs schedules of the local task sections in order to maximize the total utility gained and to minimize the number of task sections that fail to complete within the local termination time. Algorithm 1 describes the construction of local schedule by GBTS. $TL_j = \{t_1, t_2, \ldots, t_n\}$ is the list of tasks to be scheduled.

**Algorithm 1** GBTS Local Scheduling Algorithm

```
t := t_curr
Sch_tmp = sortbyUCR(TL_j)
For all t_k ∈ Sch_tmp in descending order of UCR
    If t_k.UCR > 0
        Copy Sch into Sch_tent: Sch_tent = Sch
        Insert t_k into Sch_tent at its termination time position t_k.TER
If feasible Sch_tent
    Sch := Sch_tent
Else
    Remove t_ik from Sch_tmp
     Break
t_exe := headof(Sch)
Return t_exe
```

## 3.2 Global Phase

We first describe the communication on expander graphs using gossip and then describe the global phase of the GBTS algorithm.

### 3.2.1 Gossiping Using Expander Graph

In this section, we describe the procedure used for controlling the message complexity in our algorithm. The communication graphs used in our algorithm are the conceptual data structures that limit communication patterns during gossiping in our algorithm.

The communication graph consists of all the nodes in the graph, where each node represents a processor and edges represent interconnection between processors. Each processor p in the graph can send a message to any other processor q, which is a neighbor of p in the communication graph. A crash can occur at any step of the computation. When crashes occur, the nodes representing the faulty processors are removed from the graph. So to maintain progress in communication the neighborhood of the functional processors changes dynamically. To ensure communication among the remaining nodes, the network must contain at least one good connected component, so that the messages can reach as many nodes as possible.

Expander graphs are the families of sparse graphs that have strong connectivity properties measured using edge or vertex expansion. We use the following notations and terminology to define expander graphs as follows:

**Definition 1** Expanding graphs [7]—A connected graph is called expanding if it maintains the following properties:

- P1 (Vertex Expansion Property): For any connected $S \subseteq V(t)$, where $3 \leq |S| \leq C_\alpha n/d (0 \leq C_\alpha \leq d/2)$, we have $|\Gamma(S) \backslash S| \geq C_\beta d|S| (C_\beta \in (0, 1))$
- P2 (Edge Expansion Property): Number of vertices in $S^c$ having at least $C_\delta d(|S| \backslash n)(C_\delta \in (0, 1)$ neighbors in S is less than or equal to $|S|^c - \frac{C_w n^2}{d|S|}(C_w > 0)$
- P3 (Regularity Property): The degree of the vertices $d = \Omega(\Delta)$. If degree d is equal to $\omega(\log n)$, then $d = O(\delta)$.

The first property states that connected sets have neighbors that are roughly larger than the set itself. This property ensures that an information dissemination protocol does not end in a small set but spreads to as many nodes as possible. The second property ensures that in an information dissemination process, a large number of uninformed vertices have an adequate number of informed neighbors. The last property guarantees regularity of the graph. We require our graph to have the average degree as logarithmic or have the minimum, average, and the maximum degree of the same order of magnitude. Expander graphs are graphs that satisfy all the above three properties of Definition 1. A graph having higher expansion parameters has higher resilience than a graph having lower expansion parameters.

A random walk on the graph G(t) with the states represented by V(t) and edges represented by E(t) is defined as a nonnegative $n \times n$ transition probability matrix $P = [P_{ij}]$, where $P_{ij}$ is the probability of transition from node i to node j [8]. Our motivation for using the expander graph to define the transition probability matrix is that a random walk on an expander graph converges very fast.

The energy of a node is defined similarly to the expansion factor for a node in a graph [9]. The energy of a node in the graph G(t) is defined as follows.

**Definition 2** Energy of a node ($E_x^t$)—The energy of a node x at time t is defined as the number of existing edges among the first neighbors of x divided by the number of edges that could possibly exist between the first-neighborhood of x. Formally, the energy of a node is given by

$$E_x^t = \begin{cases} \frac{2N_x(t)}{\beta_x(t)(\beta_x(t)-1)} & if \left|\Gamma_x^1(t) > 1\right| \\ 0 & \text{otherwise} \end{cases} \tag{13}$$

Let $\Gamma_G(x)$ be the set of neighbors of a node x in the graph G. A random walk from a node x to its neighbor y moves from a region of low energy to high energy. The probability transition matrix for such a walk is given by

$$P = \begin{cases} (1 - \frac{1}{\beta_x(t)}) & \text{if } E_y^t > E_x^t \\ \frac{1}{\beta_x(t)} & \text{if } E_y^t < E_x^t \\ 0 & \text{otherwise} \end{cases} \quad (14)$$

Each node x transmits with higher probability to neighbors having higher energy. This means it will have more neighbors to choose from in the next step. A crucial factor of our approach is that each node x only have to compute the energy of its immediate neighbors to calculate the transition probability matrix. So this approach is totally distributed and decentralized and hence scalable.

The gossiping process starting from an initiator node u selects a node from its neighbor according to the probability matrix P and forwards the gossip message. The time interval at the beginning of which nodes send out messages is called the gossiping time interval (also called the round or the cycle) denoted by r. The messages are believed to arrive at their destination at the end of the time interval. The nodes that are informed in the network at the end of gossiping time interval r are called informed nodes denoted by $I_r$. Initially, $I_0 = \{u\}$, $I_{r+1} = I_r \cup H_r(I_r)$, where $H_r(I_r)$ are the neighbors of the nodes in $I_r$. The nodes that are not informed in the network at the end of gossiping time interval r are called the uninformed nodes $U_r$. Initially, at time r=0 $U_0 = n - 1$.

### 3.2.2 Protocol Description

At each beginning of a gossip round, a node $N_i$ containing the present section of the task forwards a gossip message along its neighbors. After forwarding the message, $N_i$ waits for a reply until a definite deadline D, which is less than or equal to the task end to end completion time. Each node on receiving the gossip message will, in turn, forward the message along the neighbors in the expander graph, thus the gossip messages spread as fast as possible among the nodes in the network. Each node forwards the message until the node with the requested object is found. When the node containing the object receives the gossip message, it forwards a reply to $N_i$. If the node does not receive a reply within the deadline, then the task section $t_{ik}$ is aborted and messages are sent to all the previous upstream nodes informing them about the decision. Otherwise, if the node containing the object is found, then that node becomes the next head node of the task. Algorithm 2 describes the global phase scheduling of GBTS protocol.

**Algorithm 2** GBTS Global Scheduling Algorithm

**Local Phase at node $N_i$**
$t := t_{curr}$
Execute task section $t_k$
Decide the next task section $t_{k+1}$ and the object it requires
Calculate deadline for the next task section $D_{next}$
$D_{next} = t + ER_{ik} + \frac{SLACK_i^{Total}}{NS-1}$
Forwards gossip message along the edges selected with probability P.
Wait till $D_{next}$
**If** $ack = true$ from $N_k$ and $D_{next}$ not reached
    Send task to $N_k$
**Else If** $D_{next}$ reached
    Abort $t_k$
**Remote Phase at each intermediate node $N_k$ on receiving the gossip message**
**If** $noofgossipmessages \leq 1$
    Forward gossip(msg)
**Remote Phase at node $N_j$ holding object O required for next task section**
Upon receiving gossip message
**If** $accept = T_i.schedule$
    Send ack to $N_i$
    Execute task section.

## 4 Analysis of the GBTS Protocol

**Theorem 1** *Let u be an arbitrary initiator node of the expander graph. The information dissemination process using gossip starting from u on the expander graph informs all other nodes within O(log n) rounds.*

*Proof* Our analysis consists of several phases and each phase is made up of many steps. Each phase considers information dissemination from vertices $H_{r-1} \subseteq I_r$ that have been contacted in the previous phase but has not spread the information yet.

**Phase 1**: Let u be the initiator node at round r = 0. From the vertex expansion property of expansion graphs, we know that the neighborhood $\Gamma_u^1(r)$ grows exponentially at each round r. If $d = \omega(logn)$, then it is sufficient to inform only the vertices in $\Gamma_u^1(r)$. Else if $\Gamma_u^1(r-1)$ is known at round r – 1, then we can inform $\Gamma_u^1(r)$ in at most $\Delta$ steps. So for some constant $C > 0$, after r = log n rounds, $H_r \geq Clogn$ and $I_r = O(H_r)$.

**Phase 2**: We now have a set $H_r$ of size at least log n. In this phase, we aim to inform at least *n/d* vertices. From Property 1 of Definition 1, we can write that $C_2 logn \leq I_r \leq C_\alpha n/d$. Property 1 ensures that given a set of informed vertices satisfying the vertex expansion property within a constant number of steps k, the set of informed vertices increases by a factor strictly greater than one. So for a constant $C_1 > 0$, $|I_{(r+k)}| \geq C_1 |I_r|$. Thus, $log_{(C1)}(I_r) = log_{C1}(C_\alpha n/d) = O(logn)$ steps are required to reach O(n/d) uninformed vertices.

**Phase 3**: This phase aims at informing a linear number of vertices. We know from the edge expansion property of expander graph that there exist a large number of uninformed nodes that have their neighbors in $I_r$. So from phase 1 and 2, we can say that applying phase 3, log n times ensure that a linear number of vertices are informed. Continuing in this way and combining all the phases, we get $|Ir + O(logn)| \approx n$.

**Theorem 2** *The number of messages generated by the global phase of GBTS protocol is O(nloglogn) where n is the number of nodes in the network.*

*Proof* Let $d = logn - 1$. (Property 3) The initiator node selects nodes from its neighbors whose energy is greater than its own energy and forwards the gossip message to them. We calculate the expected number of nodes that are contacted by the initiator node (In the worst case, all d nodes are sent the message) as follows: The probability that exactly i nodes are sent, the gossip message will be $\theta(1/(i+1)i)$. Hence the expected number of nodes that the initiator sends the gossip message to is $\sum_{i=1}^{d} \theta(1/(i+1)i) = O(\log d)$ and so the expected number of messages is $O(logd)$. So by linearity of expectation, we can write that the expected number of gossip messages that are sent by all the nodes in the network is $O(nlogd) = (nloglogn)$.

## 5 Conclusion and Future Work

In this paper, we present a distributed scheduling algorithm using gossip for real-time tasks in large-scale unreliable networks. The main goal of GBTS is to lower the message overhead that is incurred while propagating task scheduling parameters and determining the next node for execution. As gossip-based protocols incur high message complexity, we propose to use an expander graph for controlling the message complexity of gossip. We prove that gossiping using expander graph significantly improves the message complexity of our algorithm. Future work focuses on experimentally analyzing our work and comparing it with other existing protocols.

## References


1. Han, K., et al.: Exploiting slack for scheduling dependent, distributable real-time threads in mobile ad hoc networks. In: RTNS, March 2007
2. Bettati, R.: End-to-End scheduling to meet deadlines in distributed systems. Ph.D. thesis. UIUC (1994)
3. Han, K., Ravindran, B., Jensen, E.D.: RTG-L: dependably scheduling real-timedistributable threads in large-scale, unreliable networks. In: Proceedings of IEEE Pacific Rim International Symposium on Dependable Computing (PRDC) (2007)
4. Zhang, B., Han, K., Ravindran, B., Jensen, E.D.: RTQG: Real-time quorumbased gossip protocol for unreliable networks. In: Proceedings of The Third International Conference on Availability, Reliability and Security (2008)
5. Jensen, E.D., et al.: A time-driven scheduling model for real-time systems. In: RTSS, pp. 112–122, Dec 1985

6. Kasprzyk, R.: Diffusion in networks. J. Telecommun. Inf. Technol. 2 (2012)
7. Doerr, B., Friedrich, T., Sauerwald, T.: Quasirandom rumor spreading: expanders, push vs. pull, and robustness. In: Albers, S. et al. ICALP 2009, Part 1. LNCS, vol. 5555, pp. 366–377. Springer, Heidelberg (2009)
8. Shah, D.: Gossip algorithms. In: Foundations and Trends in Networking, vol. 3, no. 1, pp. 1–125 (2009). https://doi.org/10.1561/1300000014
9. Wijetunge, U., Perreau, S., Pollok, A.: Distributed stochastic routing optimization using expander graph theory. In: IEEE Australian Communication Theory Workshop AusCTW (2011)

# Exact Algorithm for L(2, 1) Labeling of Cartesian Product Between Complete Bipartite Graph and Path

**Sumonta Ghosh and Anita Pal**

**Abstract** Graph labeling problem put nonnegative integers to the vertex with some restrictions. $L(h, k)$ labeling is one kind of graph labeling where adjacent nodes get the value difference by at least $h$ and the nodes which are at 2 distance apart get value differ by at least $k$, which has major application in radio frequency assignment, where assignment of frequency to each node of radio station in such a way that adjacent station get frequency which does not create any interference. Robert in 1988 gives the idea of frequency assignment problem with the restriction "close" and "very close", where "close" node received frequency that is different and "very close" node received frequency is two or more apart, which gives the direction to introduce $L(2, 1)$ labeling. $L(2, 1)$ labeling is a special case of $L(h, k)$ labeling where the value of $h$ is 2 and value of $k$ is 1. In $L(2, 1)$ labeling, the difference of label is at least 2 for the vertices which are at distance one apart and label difference is at least 1 for the vertices which are at distance two apart. The difference between minimum and maximum label of $L(2, 1)$ labeling of the graph $G = (V, E)$ is denoted by $\lambda_{2,1}(G)$. Here, we propose a polynomial time algorithm to label the graph obtained by the Cartesian product between complete bipartite graph and path. We design the algorithm in such a way that gives exact $L(2, 1)$ labeling of the graph $G = (K_{m,n} \times P_r)$ for the bound of $m, n > 5$ and which is $\lambda_{2,1}(G) = m + n$. Our proposed algorithm successfully follow the conjecture of Griggs and Yeh. Finally, we have shown that $L(2, 1)$ labeling of the above graph can be solved in polynomial time for some bound.

**Keywords** Cartesian product · $L(h, k)$ labeling · $L(2, 1)$ labeling
Complete bipartite graph · Path

---

S. Ghosh · A. Pal (✉)
National Institute of Technology Durgapur, Durgapur 713209, West Bengal, India
e-mail: anita.buie@gmail.com

S. Ghosh
e-mail: mesumonta@gmail.com



R. Chaki et al. (eds.), *Advanced Computing and Systems for Security*,
Advances in Intelligent Systems and Computing 897,
https://doi.org/10.1007/978-981-13-3250-0_2

# 1 Introduction

Labeling technique of graph has become very useful in the domain of applied mathematics. So, many applications like conflict resolution in social psychology, electrical circuit theory, radar location code, missile guidance code, design communication network addressing system, etc. $L(h, k)$ labeling problem has major application in channel assignment, where assignment of channel to each node of a station in such a way that adjacent station get channel which does not create any interference.

Robert in 1988 gives the idea of frequency (channel assignment problem also known as frequency assignment problem) assignment problem with the restriction "close" and "very close", where "close" node received frequency that is unique and "very close" node received frequency is two or more apart, which lead to the introduction of $L(2, 1)$ labeling. This problem is just a vertex coloring problem and here color is replaced by nonnegative integer, by which we can label the vertex of a graph $G = (V, E)$ where $V$ and $E$ represents the set of vertices the set of edges. $L(2, 1)$ labeling impose the restriction that, if $d(x, y) = 2, \forall x, y \in V$, i.e., for "close" node frequency will be at least 1 apart and if $d(x, y) = 1, \forall x, y \in V$, i.e., for "very close" node frequency will be at least 2 apart.

The graph $G = (V, E)$ has different bound of $\lambda_{2,1}(G)$, which is known in terms of $\Delta$, $\omega(G)$, and $\chi(G)$. $\Delta$ denotes the maximum degree of the graph $G$, where $\omega(G)$ and $\chi(G)$ denote the size of the maximum clique and chromatic number of the graph $G$, respectively. Some results are like the graph $K_{1,\Delta}$ stable lower bound is $\Delta + 1$, Griggs and Yeh [1] gives the explanation that a graph required $\Delta^2 - \Delta$ span, later they also prove that $\lambda_{2,1}(G) \leq \Delta^2 + 2\Delta$. Gonclaves [2] tuned the bound to $\lambda_{2,1}(G) \leq \Delta^2 + 2\,\Delta - 2$ and later, the bound is improved to $\lambda_{2,1}(G) \leq \Delta^2 + 2\,\Delta\ - 3$ for 3−connected graph. Chang and Kuo [3] furthur improved the bound. The conjecture of Griggs and Yeh [1] helps to stabilized it for the graph of diameter 2, then the bound $\lambda_{2,1}(G) \leq \Delta^2$ has been improve by Chang and Kuo [3].

**Conjecture 1** For any graph $G = (V, E)$ with maximum degree $\Delta \geq 2$, $\lambda_{2,1}(G) \leq \Delta^2$.

The above conjecture of Griggs and Yeh [1] worked for the set of graphs, like path [1], wheel [1], cycle [1], trees [1, 3, 4], co-graphs [3], interval graphs [3], chordal graphs [5], permutation graph [6, 7], etc. The bound $\lambda_{2,1}(G)$ can be computed systematically for some graphs, like cycle, path, tree [1, 3, 4]. For some graphs $G = (V, E)$, like path, cycle, complete bipartite graph, tree, star, bi-star, complete graph $\lambda_{2,1}(G)$ is computable in polynomial time, but some other class of graphs is also there, namely interval graph [3], circular arc graph [8], chordal graph [5], etc. $\lambda_{2,1}(G)$ may not be computed in polynomial time, complexity of such graphs are either NP-complete or NP-hard. The study of $\lambda_{2,1}(G)$ of Cartesian product between paths, cycles, complete graphs and between paths and cycles has already been done [9–19].

When labeling problem gives exact labeling, it become very interesting. Exact $L(2, 1)$-labeling problemis acquired by many researchers and they also have done some useful work on this, and still it is very much criticized and challenging work. Simple graph like path, cycle, complete bipartite graph, and tree can give exact $L(2, 1)$-labeling, but for complex graph structure like Cartesian product between different graphs is not as smooth as simple graphs. Here, we concentrate on exact $L(2, 1)$-labeling of Cartesian product between complete bipartite graph and path.

Faster development in communication system and due to rapid digitalization of everything, we enter into a wider, complex, and hybridization network structure which leads to increase the number of node or station. Enormous number of nodes in any communication system may lead to collision, so to avoid collision, we have to impose restriction. Increasing of label(frequency) may lead to high-cost factor which may affect the feasibility of maintaining such complex network. So, we partly introduce exact $L(2, 1)$-labeling to remove high-cost factor and reducing failure to maintain reliability of the system.

The remaining part of the paper is organized in the way that Sect. 2 contains some preliminaries and definition, Sect. 3 presents algorithms, analysis of algorithm, and Lemma's to study Griggs and Yeh [1] conjecture followed by conclusion.

## 2 Preliminaries

**Definition 1** A graph $G$ is called a complete bipartite graph if its vertices can be partitioned into two subsets $V_1$ and $V_2$ such that no edges has both end points in the same subset, and each vertex of $V_1(V_2)$ is connected with all vertices of $V_2(V_1)$. Here $V_1 = \{X_{11}, X_{12}, \ldots, X_{1m}\}$ contains $m$ vertices and $V_2 = \{Y_{11}, Y_{12}, \ldots, Y_{1n}\}$ contains $n$ vertices.

A complete bipartite graph with $|V_1| = m$ and $|V_2| = n$ is denoted by $K_{m,n}$.

**Definition 2** $L(2, 1)$-labeling is the special case of $L(h, k)$ labeling where the value of $h$ is equal to 2 and the value of $k$ is equal to 1. $L(2, 1)$ labeling of a graph $G = (V, E)$, where $V$ represents the vertex set $E$ represent the edge set and $f$ is a function whose mapping from the set of vertices $V$ to the set of positive integer such that $|f(x) - f(y)| \geq 2$ if distance $d(x, y) = 1$ and $|f(x) - f(y)| \geq 1$ if distance $d(x, y) = 2$. Difference between maximum label and minimum label of $L(2, 1)$-labeling $f$ of $G$ denoted by $\lambda_{2,1}(G)$.

**Definition 3** Cartesian product is denoted by $G \times H$ where $G = (V, E)$ and $H = (V', E')$ be the graph two graphs. Which is defined by taking Cartesian product between two set of vertices $V(G) \times V'(H)$, where $(u, u')$ and $(v, v')$ are the order pair of the Cartesian product will be directly connected in $G \times H$ if and only if either

1. $u = v$ and $u'$ is directly connected with $v'$ in $H$, or
2. $u' = v'$ and $u$ is directly connected with $v$ in $G$.

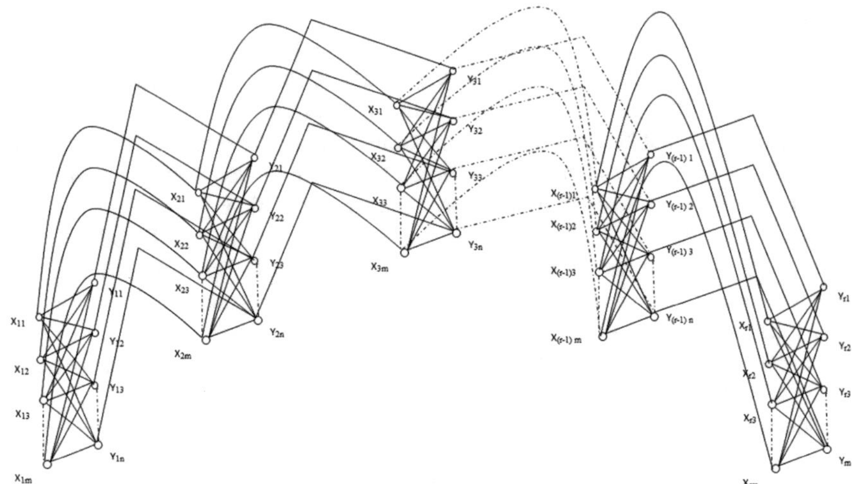

**Fig. 1** The graph $K_{m,n} \times P_r$

For Cartesian product between $K_{m,n} \times P_r$, we have to draw the graph $K_{m,n}$, $r$ times. Here, each $K_{m,n}$ has two set of vertices $X$, $Y$ where $|X| = m$ and $|Y| = n$. Each set of vertices of $K_{m,n}$ for $r$ copies is represented by $(X_1, Y_1)$, $(X_2, Y_2)$, $(X_3, Y_3)$, ..., $(X_r, Y_r)$, where each $X_i = \{x_{i1}, x_{i2}, x_{i3}, \ldots, x_{im}\}$ and $Y_i = \{y_{i1}, y_{i2}, y_{i3}, \ldots, y_{in}\}$. Consider the graph $G = (K_{m,n} \times P_r)$, where $V = \bigcup_{i=1}^{r} v_i$ and $v_i = (X_i, Y_i)$, where $X_i = \{x_{i1}, x_{i2}, x_{i3}, \ldots, x_{im}\}$ and $Y_i = \{y_{i1}, y_{i2}, y_{i3}, \ldots, y_{in}\}$ for each $i = 1, 2, 3, \ldots, r$. Two vertices are to be connected by the following.

1. $x_{i_1 j}$ and $x_{i_2 k}$ will be connected if $j = k$ and $|i_1 - i_2| = 1$.
2. $y_{i_1 p}$ and $y_{i_2 q}$ will be connected if $p = q$ and $|i_1 - i_2| = 1$.

and $(x_{ij}, y_{ip}) \in E$, i.e., $d(x_{ij}, y_{ip}) = 1$ for $i = 1, 2, 3, \ldots, r$, $j = 1, 2, 3, \ldots, m$ and $p = 1, 2, 3, \ldots, n$. Again $d(x_{ij}, x_{(i+1)j}) = 1$ for $i = 1, 2, 3, \ldots, r$, $j = 1, 2, 3, \ldots, m$ and $d(y_{ip}, y_{(i+1)p}) = 1$ for $i = 1, 2, 3, \ldots, r$, $p = 1, 2, 3, \ldots, n$ (Fig. 1).

**Lemma 1** *Let $\triangle$ be the degree of the graph $K_{m,n} \times P_r$, then*

$$\triangle = \begin{cases} m+1 & for\ m > n\ and\ r = 2 \\ m+2 & for\ m > n\ and\ r > 2 \\ m+1 & for\ m = n\ and\ r = 2 \\ m+2 & for\ m = n\ and\ r > 2 \end{cases} \quad (1)$$

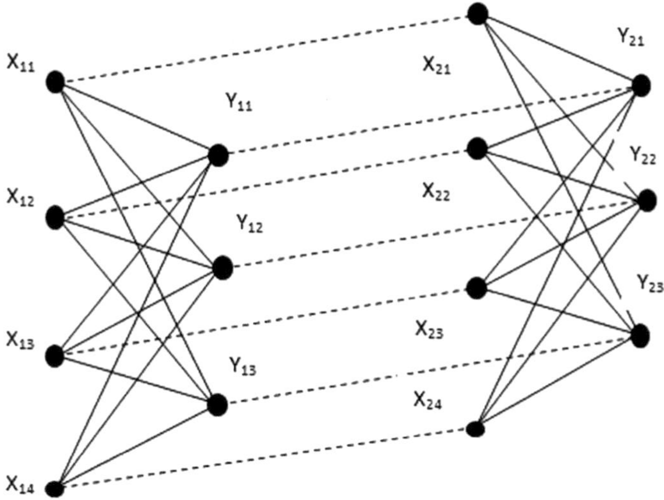


**Fig. 2** Cartesian product between $K_{m,n}$ and $P_r$ for $m > n$ and $r = 2$

*Proof* Let $G = K_{m,n} \times P_2$. If $m > n$ then $K_{m,n}$ has the maximum degree $m$. As per definition, the vertex $x_{1j}$ connected with vertex $x_{2j}$, $j = 1, 2, 3, \ldots, m$ and $y_{1i}$ is connected with $y_{2i}$, $i = 1, 2, 3, \ldots, n$. Therefore, only one degree of each vertex will increase in $K_{m,n} \times P_2$. Hence, the value of $\Delta$ is $m + 1$.

The proof of other cases is similar.

## 3 Labeling of Cartesian Product Between Complete Bipartite Graph and Path

Introduction section is already enriched with various types of labeling like intersection graphs, Cartesian product of some graphs with their bounds $\lambda_{2,1}(G)$ in the form of $\Delta$, and number of vertices. Now, we discussed about the $L(2, 1)$ labeling of path $P_r$ and exact algorithmic idea of $L(2, 1)$ labeling of Cartesian product between complete bipartite graph and path followed by analysis of algorithm. We also show that the algorithm also follows the Griggs and Yeh conjecture (Figs. 2 and 3).

In this paper, we consider $G = K_{m,n} \times P_r$, where $m, n > 5$, for exact $L(2, 1)$ labeling. We give the name exact $L(2, 1)$ labeling because number of labels require to label a complete bipartite graph $K_{m,n}$, for $m, n > 5$ by $L(2, 1)$ labeling is equal to the number of labels require to label the graph which we obtained by doing Cartesian product between complete bipartite graph and path by same labeling scheme. We consider the restriction $m, n > 5$ because it always maintain the exact property discussed previously. For $m, n \leq 5$, we are able to label by $L(2, 1)$ labeling but it failed to attend the exact labeling scheme.

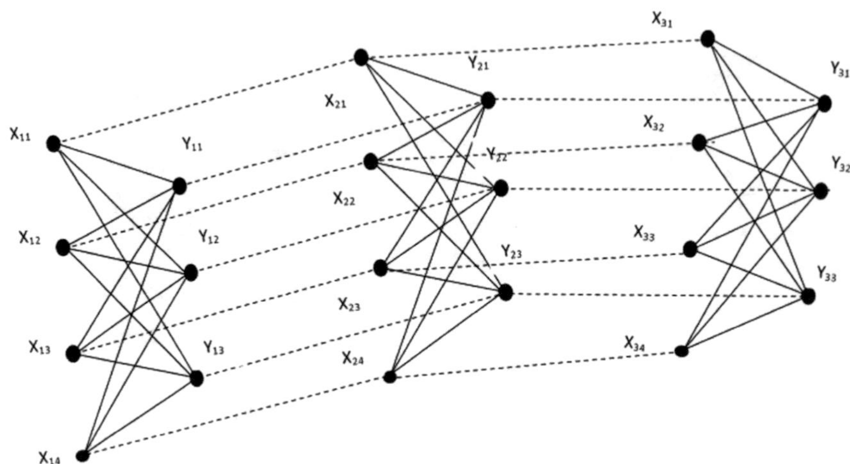


**Fig. 3** Cartesian product between $K_{m,n}$ and $P_r$ for $m > n$ and $r = 3$

## 3.1 Algorithm L21P

We use the algorithm L21P for labeling the path $P_r$ of length $r$ by $L(2, 1)$ labeling. considering the vertices $v_0, v_1, v_2, \ldots, v_{r-1}$ of the path $P_r$ where $v_i$ and $v_{i+1}$ is adjacent. To label the path $P_r$, we consider one array $V[r]$ to store the labeling value. The technique of labeling of path $P_r$ are given below.

## 3.2 Algorithm L21CBG

To label the complete bipartite graph $K_{m,n}$ with vertex set $X$ and $Y$. We consider two arrays $X[m]$ and $Y[n]$ to store the labeling value of $K_{m,n}$.

**Theorem 1** *For complete bipartite graph $G = K_{m,n}$, $\lambda_{2,1}(K_{m,n}) = m + n$.*

---

**Algorithm 1** $L(2, 1)$-Labeling of Path (L21P)

---

**Input:** $G = P_r$.
**Output:** labeled graph $G = P_r$.
**Initialize:** Array $V[r]$.
**Step 1:** loop $i = 0$ to $(r - 1)$.
**Step 2:** If $i$ is divisible by 5 then $V[i] = f(v_i) = 2$.
**Step 3:** If $(i - 1)$ is divisible by 5 then $V[i] = f(v_i) = 0$.
**Step 4:** If $(i - 2)$ is divisible by 5 then $V[i] = f(v_i) = 3$.
**Step 5:** If $(i - 3)$ is divisible by 5 then $V[i] = f(v_i) = 1$.
**Step 6:** If $(i - 4)$ is divisible by 5 then $V[i] = f(v_i) = 4$.
end of loop $i$.
**Stop.**

---

**Algorithm 2** $L(2, 1)$-Labeling of Complete Bipartite Graph (L21CBG)

**Input:** $G = K_{m,n}$.
**Output:** labeled graph $G = K_{m,n}$.
**Initialize:** $c = 0$.
**Step 1** loop $i = 0$ to $(r - 1)$.
**Step 2** $X[i] = c, c = c + 1$.
end of loop $i$.
**Step 3** $c = c + 2$
**Step 4** loop $j = 1$ to $n$
**Step 5** $Y_k[j] = c, c = c + 1$.
end of loop $j$.
Stop.

*Proof* Let us consider the complete bipartite graph $K_{m,n}$ with two set of vertices $X = \{x_1, x_2, x_3, \ldots, x_m\}$ and $Y = \{y_1, y_2, y_3, \ldots, y_n\}$, where $|X| = m$ and $|Y| = n$. It is clear that the vertices within a set is not connected, so each vertex in a particular set is at distance 2, whereas any two vertices from different sets are at distance 1. If we start labeling the vertex set $X$ with 0, i.e., $f(x_1) = 0$, we can increase label by 1 for the next vertex because all the vertices of set $X$ are at distance two apart. So, we can continue with $f(x_2) = 1$, $f(x_3) = 2$, $f(x_4) = 3$ similarly $f(x_m) = (m - 1)$. We can start label the set $Y$ by the label $(m - 1) + 2$, i.e., $m + 1$. So $f(y_1) = m + 1$,$f(y_2) = m + 2$, $f(y_3) = m + 3$, $f(y_4) = m + 4$ similarly $f(y_n) = m + n$. So $\lambda_{2,1}(K_{m,n}) = m + n$.

## 3.3 Algorithm EL21LCP

We consider the graph Cartesian product between complete bipartite graph and path, i.e., $G = K_{m,n} \times P_r$. This can also be incorporate in computer memory, for easy to understand we use two array for each copy of $K_{m,n}$, these are $X_k[i]$ and $Y_k[j]$. According to the Fig. 4, we consider the path form by the array elements $X_k[0]$, $k = 1, 2, 3, \ldots, r$, which is the first vertex of set $X$ for each copy of $K_{m,n}$. Another array $P_{array}[r]$ is consider for storing the $L(2, 1)$ labeling of path form by the first vertex of set $X$ for each copy of $K_{m,n}$, which is labeled by the Algorithm 1. For the Algorithm 2, we consider two variable $maxX$ and $maxY$ to store the maximum label use by $X_1[m]$ and $Y_1[n]$, respectively.

**Correctness proof of EL21LCC is given below**

**Theorem 2** *Algorithm EL21LCC exactly label the graph* $G = (K_{m,n} \times P_r)$.

*Proof* According to the Theorem 1, we know that $\lambda_{2,1}(K_{m,n}) = m + n$. It is clear that if we label complete bipartite graph $K_{m,n}$ starting by 0, then $maxX = m - 1$ and $maxY = m + n$. Now, we consider the Cartesian product between complete bipartite

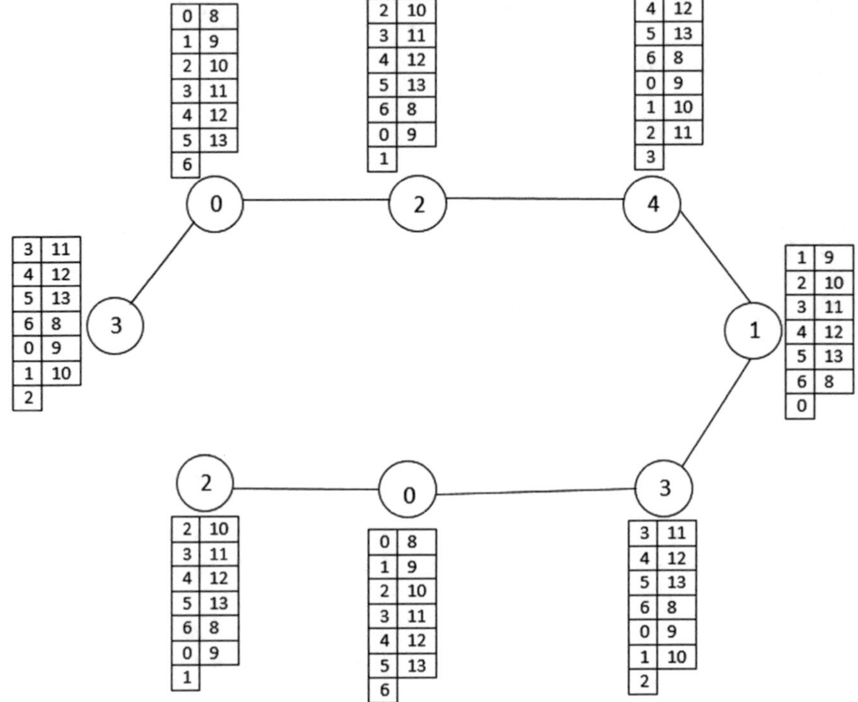


**Fig. 4** Exact $L(2, 1)$-Labeling of the graph $K_{7,6} \times P_8$

---

**Algorithm 3** Algorithm for $L(2, 1)$-Labeling of Cartesian Product Between Complete Bipartite Graph and Path (EL21LCP)

---

**Input:** $G = (K_{m,n} \times P_r)$.
**Output:**labeled graph $G = (K_{m,n} \times P_r)$.
**Initialize:** $maxX = m - 1, maxY = m + n, Y_{k-1}[1] = maxX + 2, X_{k-1}[1] = 0$.
**Step 1:** For $L(2, 1)$ labeling of the path $P_r$ call the algorithm $L21P$
and store the value in the array $P_{array}[r]$.
**Step 1:** loop $k = 1$ to $r$)
**Step 2:** $p = P_{array}[k]$
**Step 3:** loop $i = 1$ to $m$
**Step 4:** if $(p > maxX)$ then $p = 0$
end of if.
**Step 5:** $X_k[i] = p; p = p + 1$
end of loop $i$.
**Step 6:** $q = Y_{k-1}[1] + (X_k[1] - X_{k-1}[1])$
**Step 7:** loop $j = 1$ to $n$
**Step 8:** if $(q > maxY)$ then $q = maxX + 2$
end of if.
**Step 9:** $Y_k[j] = q, q = q + 1$.
end of loop $j$.
end of loop $k$.
Stop.

---

graph and path and we got the graph $G = (K_{m,n} \times P_r)$, where $m, n > 5$ (see Fig. 4). From the graph $G = K_{m,n} \times P_r$, we get $r$ copies of $K_{m,n}$ and we consider for each copy of $K_{m,n}$ two array $X$ and $Y$. We consider array $P_{array}$ to label the path, which is formed by the each first vertex of set $X$. The array $P_{array}[r]$, first element is the label for the first vertex of vertex set $X$ for first copy of $K_{m,n}$, i.e., $X_1[1] = P_{array}[1]$, second element is the label for the first vertex of vertex set $X$ for second copy of $K_{m,n}$, i.e., $X_2[1] = P_{array}[2]$, similarly $r$th element is the label for the first vertex of vertex set $X$ for $r$th copy of $K_{m,n}$, i.e., $X_r[1] = P_{array}[r]$. Now, we fetch the first element of the array $P_{array}[r]$ and start labeling the first copy of $K_{m,n}$ according to Algorithm 3, next fetch the second element from the array $P_{array}$ and start labeling the second copy of $K_{m,n}$ according to Algorithm 3 just when it attends the value $maxX$ and $maxY$, we assign the label 0 and $m + 1$, respectively, to the very next vertex. A similar way can followed for the remaining step. By shuffling the existing label, we can achieve exact labeling for the graph $G = (K_{m,n} \times P_r)$ as the $K_{m,n}$, i.e., $\lambda_{2,1}(K_{m,n}) = \lambda_{2,1}(K_{m,n} \times P_r) = m + n$.

**Analysis of Algorithm EL21LCP**: Lets consider the graph $G = K_{7,6} \times P_8$ to analyze the algorithm EL21LCP. For $L(2, 1)$ labeling of the path $P_8$ call the algorithm $L21C$ and store the value in the array $P_{array}[8]$ and the corresponding labels are $P_{array} = \{2, 0, 3, 1, 4, 2, 0, 3\}$. Initially $maxX = 7 - 1 = 6$, $maxY = 7 + 6 = 13$, $Y_{k-1}[1] = maxX + 2 = 6 + 2 = 8$ and $X_{k-1}[1] = 0$. For $k = 1$, we start labeling the first copy of $K_{7,6}$, now initial value of $p = P_{array}[1] = 2$. For $i = 1$ first check $p = 6$ or not, clearly it is false then $X_1[1] = 2$ and $p = p + 1 = 1$. For $i = 2$, checking condition false and $X_1[2] = 3$, when $i = 6$ checking condition is true and assign the value $p = 0$ and last two label for first copy of set $X$ is $X_1[6] = 0$ and $X_1[7] = 1$. Now, $q = Y_{k-1}[1] + (X_k[1] - X_{k-1}[1]) = 8 + (2 - 0) = 10$, for $j = 1$ check whether $q = 13$ or not, which is false then $Y_1[1] = 10$, $q = q + 1 = 11$. for $j = 2$, checking conditions is false and $Y_1[2] = 11$, $q = q + 1 = 12$. when $j = 5$ checking condition is true and assign the value $q = maxX = 2 = 8$ and last two label for first copy of set $Y$ is $Y_1[5] = 8$ and $Y_1[6] = 9$. For next iteration $K = 2$ we start labeling the second copy of $K_{7,6}$, now value of $p = P_{array}[2] = 0$. For $i = 1$ condition checking gives false and $X_2[1] = 0$, similarly for $i = 2, 3, 4, 5, 6, 7$ corresponding label is $X_2[2] = 1$, $X_2[3] = 2$, $X_2[4] = 3$, $X_2[5] = 4$, $X_2[6] = 5$, $X_2[7] = 6$. Now, $q = Y_{k-1}[1] + (X_k[1] - X_{k-1}[1]) = 10 + (0 - 2) = 8$, for $j = 1$ check whether $q = 13$ or not, which is false and then $Y_2[1] = 8$, similarly for the value of $j = 2, 3, 4, 5, 6$ corresponding label is $Y_2[2] = 9$, $Y_2[3] = 10$, $Y_2[4] = 11$, $Y_2[5] = 12$, $Y_2[6] = 13$. Similarly, we can label rest of the copies of $K_{7,6}$ and we carefully observe that label will not increase more that 13. Actually, we design the algorithm in such a way where we shuffling the existing label of first copy of $K_{7,6}$.

**Time Complexity Analysis of Algorithm EL21LCP**: According to the algorithm EL21LCP, the first loop will work for the path length $r$ then, for each encounter, next two serial inner loops will run for $m$ times and $n$ times to label each complete bipartite graph $K_{m,n}$. So, to exact label, the whole graph that obtained by doing Cartesian product between complete bipartite graph and path is either $\bigcirc(mr)$ or $\bigcirc(nr)$ depending on the value of $m$ and $n$.

**Griggs and Yeh conjecture** Algorithm 2 successfully follow the Griggs and Yeh conjecture for $m, n > 5$.

**Case 1: Griggs and Yeh conjecture satisfy for** $m > n$ **and** $r = 2$.

**Theorem 3** *for the graph* $G = (K_{m,n} \times P_r)$ *we already shown in the Lemma 1 that maximum degree is* $m + 1$. *Now from the Theorem 1 we can know that* $\lambda_{2,1}(K_{m,n} \times P_r) = m + n$. *As we know.*
$(m + 1)^2 = m^2 + 2m + 2.$
$(m + 1)^2 > m^2 + m.$
$(m + 1)^2 > m + n$ *as* $m > n$.
*Hence the proof.*

**Case 2: Griggs and Yeh conjecture satisfy for** $n > m$ **and** $r = 2$.

**Theorem 4** *for the graph* $G = (K_{m,n} \times P_r)$ *it is clear from the Lemma 1 that maximum degree is* $n + 1$. *Now from the Theorem 1 we can know that* $\lambda_{2,1}(K_{m,n} \times P_r) = m + n$. *As we know.*
$(n + 1)^2 = n^2 + 2n + 2.$
$(n + 1)^2 > n^2 + n.$
$(n + 1)^2 > m + n$ *as* $n > m$.
*Hence the proof.*

**Case 3: Griggs and Yeh conjecture satisfy for** $m > n$ **and** $r > 2$.

**Theorem 5** *for the graph* $G = (K_{m,n} \times P_r)$ *we already shown in the Lemma 1 that maximum degree is* $m + 2$. *Now from the Theorem 1 we can know that* $\lambda_{2,1}(K_{m,n} \times P_r) = m + n$. *As we know.*
$(m + 2)^2 = m^2 + 4m + 4.$
$(m + 2)^2 > m^2 + m.$
$(m + 2)^2 > m + n$ *as* $m > n$.
*Hence the proof.*

**Case 4: Griggs and Yeh conjecture satisfy for** $n > m$ **and** $r > 2$.

**Theorem 6** *for the graph* $G = (K_{m,n} \times P_r)$ *it is clear from the Lemma 1 that maximum degree is* $n + 2$. *Now from the Theorem 1, we can know that* $\lambda_{2,1}(K_{m,n} \times P_r) = m + n$. *As we know.*
$(n + 2)^2 > n^2 + n.$
$(n + 2)^2 > m + n$ *as* $n > m$.
*Hence the proof.*

## 4 Conclusion

This work is the first attempt of $L(2, 1)$ labeling of a graph obtained by doing Cartesian product between complete bipartite graph and path. Labeling of graph becomes an interesting research domain as it is directly deal with real-world problem, $L(2, 1)$ labeling also has a great impact on different types of graph structure. A small number of graphs for which efficient algorithm is available in $L(2, 1)$ labeling. Exact $L(2, 1)$ labeling of a complex graph like $G = K_{m,n} \times P_r$ complexity with polynomial time is preferable. In this paper, we design an exact algorithm with superlinear time complexity to label a graph obtained by Cartesian product between complete bipartite graph and path, i.e., $G = (K_{m,n} \times P_r)$. But our algorithm will work for a certain bound that is for $m, n > 5$ for any $r$ and $m, n > 4$ for $r < 5$ It is an open problem for all the researchers to label a graph obtaining by Cartesian product between two simple graph complexities with polynomial time. Generating more complex structure and finding a feasible pattern to label with a restriction is highly appreciating and worthy.

**Acknowledgements** The work is supported by the Department of Science and Technology, New Delhi, India, Ref. No. SB/S4/MS: 894/14.

## References

1. Griggs, J., Yeh, R.K.: Labeling graphs with a condition at distance two. SIAM J. Discret. Math. **5**, 586–595 (1992)
2. Gonalves, D.: On the L(d, 1)-labellinng of graphs. Discret. Math. **308**, 1405–1414 (2008)
3. Chang, G.J., Kuo, D.: The L(2, 1)-labeling on graphs. SIAM J. Discrete Math. **9**, 309–316 (1996)
4. Hasunuma, T., Ishii, T., Ono, H., Uno, Y.: A linear time algorithm for L(2, 1)-labeling of trees. Lect. Notes Comput. Sci. **5757**, 35–46 (2009)
5. Sakai, D.: Labeling chordal graphs with a condition at distance two. SIAM J. Discret. Math. **7**, 133–140 (1994)
6. Paul, S., Pal, M., Pal, A.: L(2, 1)-labeling of permutation and bipartite permutation graphs. Math. Comput. Sci. https://doi.org/10.1007/s11786-014-0180-2
7. Bodlaender, H.L., Kloks, T., Tan, R.B., Leeuwen, J.V.: Approximations for -colorings of graphs. Comput. J. **47**(2), 193–204 (2004)
8. Calamoneri, T., Caminiti, S., Olariu, S., Petreschi, R.: On the L(h, k)-labeling of co-comparability graphs and circular-arc graphs. Networks **53**(1), 27–34 (2009)
9. Georges, J.P., Mauro, D.W., Stein, M.I.: Labeling products of complete graphs with a condition at distance two. SIAM J. Discrete Math. **14**, 28–35 (2000)
10. Georges, J.P., Mauro, D.W., Whittlesey, M.A.: Relating path coverings to vertex labelings with a condition at distance two. Discrete Math. **135**, 103–111 (1994)
11. Jha, P.K.: Optimal L(2, 1)-labeling of Cartesian products of cycles, with an application to independent domination. IEEE Trans. Circuits and SystI. **47**(10), 1531–1534 (2000)
12. Jha, P.K., Narayanan, A., Sood, P., Sundaram, K., Sunder, V.: On L(2, 1)-labelings of the Cartesian product of a cycle and a path. Ars Combin. 55 (2000)
13. Jha, P.K., Klavar, S., Vesel, A.: Optimal L(2, 1)-labelings of certain direct products of cycles and Cartesian products of cycles. Discrete Appl. Math. **152**, 257–265 (2005)

14. Klavar, S., Vesel, A.: Computing graph invariants on rotagraphs using dynamic algorithm approach: the case of (2, 1)-colorings and independence numbers. Discrete Appl. Math. **129**, 449–460 (2003)
15. Whittlesey, M.A., Georges, J.P., Mauro, D.W.: On the number of Qn and related graphs. SIAM J. Discrete Math. **8**, 499–506 (1995)
16. Hale, W.K.: Frequency assignment: Theory and applications. Proc. IEEE. **68**, 1497–1514 (1980)
17. Krl, D., Skrekovski, R.: A theory about channel assignment problem. SIAM J. Discret. Math. **16**, 426–437 (2003)
18. Havet, F., Reed, B., Sereni, J.S.: L(2, 1)-labeling of graphs, In: Proceedings of the 19th Annual ACMSIAM Symposium on Discrete Algorithms, SODA 2008, SIAM, pp. 621–630 (2008)
19. Schwarza, C., Troxellb, D.S.: L(2, 1)-labelings of Cartesian products of two cycles. Discret. Appl. Math. **154**, 1522–1540 (2006)

# Extraction and Classification of Blood Vessel Minutiae in the Image of a Diseased Human Retina

**Piotr Szymkowski and Khalid Saeed**

**Abstract** The work presents a created methodology for detecting minutiae in the image of the retina of a sick person's eye. The worked out algorithm is used to find areas of distribution and classification of minutiae in ill human eye retinal image. The main goal of the proposed approach is to classify all minutiae into groups based on the distance from the center of the image and the distance from the edge of the image that is closer to the blind spot. For the separation of blood vessels from the image, Otsu algorithm and background subtraction were used. To get line representation of blood vessels that helps find minutiae in images, the K3M thinning algorithm was used. The proposed algorithm shows one of the most basic solutions for finding the characteristic points in a biometric image. The last step of the presented algorithm introduces an example of minutiae classification.

**Keywords** Blood vessels · Minutiae · Eye disease · Biometrics

## 1 Introduction and State of the Art

The functioning of the human body is one of the greatest mysteries that scientists face. It has regenerative capabilities, has the ability to move and grow. The human body also has coded information in it. This information can have any form: bottoms, fingerprints on the fingers or the blood vessel system in the hand. Thanks to the unique information that is contained in our body, it is possible to recognize a person by its features.

One of the media carrying information can be the human eye. It allows you to identify a human being in two different ways.

P. Szymkowski (✉) · K. Saeed
Faculty of Computer Science, Bialystok University of Technology, Bialystok, Poland
e-mail: prszymkowski@gmail.com

K. Saeed
e-mail: k.saeed@pb.edu.pl



R. Chaki et al. (eds.), *Advanced Computing and Systems for Security*,
Advances in Intelligent Systems and Computing 897,
https://doi.org/10.1007/978-981-13-3250-0_3

Currently, biometrics provides two methods for using the eye as an access key. The first one uses the color distribution in the iris, whilst the other is the flow of blood vessels in the retina of the eye. The second one is controversial due to the large number of diseases that can affect the shape and visibility of the blood vessels in the eye socket.

The aim of the study was to create an algorithm that allows detection of blood vessels as well as the examination of the retinal image of the retina patient to determine the amount and quality of minutiae present in the image. Achieving this goal may be the beginning of further work related to the diagnosis of retinal diseases and the examination of the usefulness of biopsy images of retinal patients. The scope of work involved the processing of an image of a patient's retina, i.e., vascular segmentation, binarization of the image and thinning. Enables further operations such as calculating detected minutiae, determining their type, location, and distance from the center of the image.

Researchers have selected photos of retinas for people suffering from one of four ailments: embolism, thrombosis, bloodshed, and retinal detachment.

An embolism is a situation in which the light of the artery is closed by a blood clot, air bubble, amniotic fluid, fat, or other materials transported by blood vessels. Embolism is less frequent than clots [1]. Sick people usually complain of sudden, painless loss of vision in one eye. There may also be central or paracentral glaucoma and tunnel vision [14]. During the test, retinal bleeding may be seen in the area cut off from blood vessel access. There may also be small retinal hemorrhages, narrowing of the arteries.

The embolism usually occurs at the site of the fork of the vessel. Due to the occurrence of embolism can be divided into the following:

- embolism of central branch artery,
- central artery embolism,
- embolism of the ciliary and retinal artery.

There is currently no information on the clinical efficacy of treatments for the abovementioned embolism [1].

Blood clots form inside the blood vessels. A larger size thrombosis stops in the blood vessel, causing it to lack blood flow in the vessel and thus lack of nutrients and oxygen (see Fig. 1). There are no known causes of this disease, but predispositions to it occur. Stools are most likely in the veins, where free blood flow occurs [4]. They are exposed to elderly people, people who are deprived of physical activity and people immobilized (e.g., after stroke, coma, and paresis). Those changes are often also formed during childbirth and surgery.

Blood clots in retinal vessels are characterized by petechiae, which can lead to retinal ischemia and macular edema [1]. Depending on the place of occurrence, they can be divided into the following:

- central vein thrombosis,
- branch vein thrombosis.

Depending on the type of disease, symptoms and treatments are different [1].

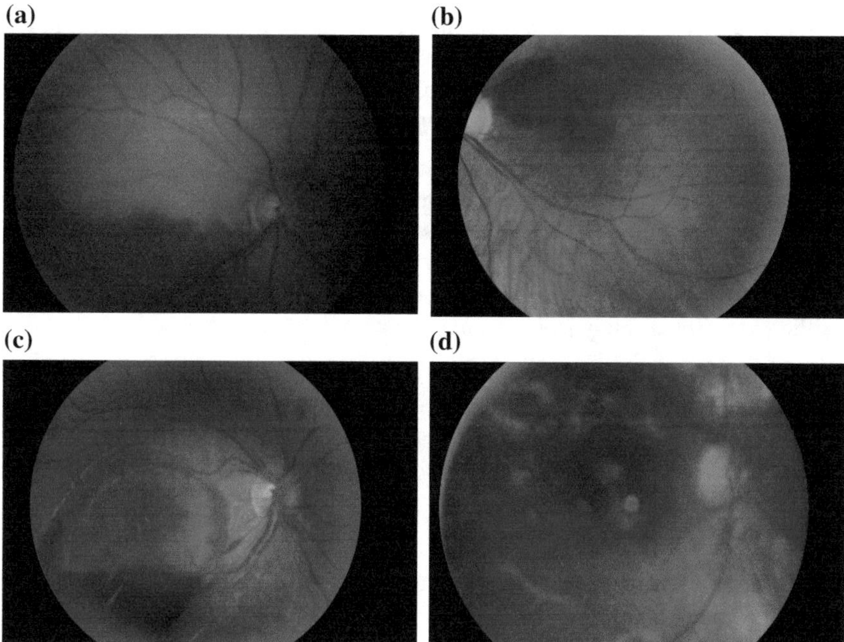

**Fig. 1** Images of retinas with diseases: embolism (**a**), thrombosis (**b**), hemorrhage (**c**) and retinal detachment (**d**)

Thrombosis of the central vein is characterized by visible changes within the closed vessel, which results in reduced visual acuity. There were no racial trends related to the occurrence of this condition. One of the most common risk factors is hypertension. The person, reporting this variant of the disease, is complaining about the distortion of vision and gravity. Treatment is done by laser or by injection [14].

The risk factors for venous thromboembolism include age over 50, tobacco smoking, hypertension, diabetes mellitus, and vasculitis [1]. Eczema can occur on a larger area of the retina. There are two variations of this condition:

- Without ischemia, which is characterized by a milder form, where there is fewer episodes and better than the type of ischemia of visual acuity.
- With ischemia worse visual acuity and a significant number of blistering pimples is characterized.

Treatment similar to that of central vein thrombosis consists of injection, laser treatment, and the addition of steroid medications [4].

Hemorrhage usually occurs as a result of pressure stroke inside the eye vein. This leads to the blood entering space between two layers of the retina [14]. After a long resting period, there is a division into the red blood cells falling to the bottom and floating over other blood components.

In the case of a sputum, it is recommended to observe changes, sometimes it is also recommended to pierce the change with a laser [1].

Retinal detachment is a situation in which the neurosensory layer separates from the retinal pigment layer behind it (see Fig. 1). There are three types of detections:

- Traction—detachment caused by shrinking glassy retinal membrane.
- Extreme—can be caused by tumors, hypertension, or inflammation. It depends on the penetration of the fluid from the vascular layer into the parietal space through the damaged iris.
- Pre-rupture—is caused by the damage of the neurosensory layer, whereby the fluid filling the vitreous body exits, causing decay.

Sometimes, the three types mentioned above can occur simultaneously [14]. The most common causes of detachment are all kinds of head injuries, eye surgery, diabetes mellitus, myopia. The risk of illness increases with age and in the case of prior family illnesses [1]. Depending on the type of detachment may occupy different sizes of areas (see Fig. 1).

Symptoms can be the following:

- painless decrease in visual acuity in one eye,
- disturbed field of view,
- Retinal projection is visible in the study.

In trabecular retinal detachment, surgical intervention is often required, whereas in the other two types (exudative, pruritus) the effects depend on the occurrence, magnitude of lesions, and their position relative to the spot [4].

One of the most important operations on the retinal image is segmentation of vessels. It allows to determine the location of the disease and the type of illness. Many variants of operations have been created to highlight the characteristics sought.

The first method uses the c-means method to detect blood vessels. To remove the noise, a 5 by 5 averaging filter is used. The next step is to use a filter that can detect the edge. The most commonly used filter for this purpose is Laplacian [5]. Depending on its appearance, it can detect edges only vertical or horizontal, or detect those that face the bevel.

The next step is to create grayscale charts to obtain such image information as energy, uniformity, contrast, and correlation. On these data, the fuzzy c-means algorithm is used to search for blood vessels.

Another method used for blood vessels search is the Simple Linear Iterative Clustering (SLIC) algorithm. It uses the same algorithm as the previous k-mediated algorithm, the difference being that the method in question uses it in a five-dimensional space, since each element has three color components and two coordinates.

Segmentation with this method is created by creating superpixels, that is, by creating pixel groups containing similar values. The image after the operation looks like a mosaic. Thus, by selecting the appropriate parameter, it is possible to remove the excess elements and noise [6, 10].

Another method of segmentation is the Canny algorithm. It is a procedure for detecting edges. It is conducted in three stages:

- Image smoothing with Gaussian filter,
- Suppression of nonparametric values,
- Image calibration.

The first step consists in filtering a low-pass image of a Gaussian filter to remove noise. The basis of this part of the algorithm is an automatic selection of standard deviation value that characterizes the Gaussian distribution, and hence the size of the mask used [11].

Next step is to find the center of the edge and reject the noise. Elimination of non-pixelated pixels produces a 1 pixel-thick image.

The last step is the threshold. Two thresholds are used, one of which limits values from top to bottom, allowing you to remove unnecessary edges. To allow further treatment with blood vessels, the space between vascular contours is required [13].

Another method is to use Gaussian matched filters. The key to using this algorithm is to create masks that detect piecewise linear segments of blood vessels. That method was used in other work [15] as a way to segmentation blood vessels. It is presenting as well another way to achieve a similar goal—creating feature vector.

## 2 Image Processing

In order to be able to investigate the usefulness of samples (images) for biometric purposes, an application has been developed which allows the use of all kinds of filters in order to obtain fully satisfactory results.

Low-pass filters are one of the types of contextual filters. This means that operations on a given item require and use knowledge of its neighbors. Value is generated by using the wave function [7, 9]. The filter allows you to limit or level high-frequency components and to pass low-frequency components. It is often used to remove noise in the image, that is, changes in the color of the elements in the test image created during its execution or processing [8].

The following filters have been implemented in the application (see Figs. 2 and 3):

- Mean filter is a basic low-pass filter, it equally takes into account the color measures of the examined pixel and its neighboring pixels.

**(a)**

| | | |
|---|---|---|
| 1 | 1 | 1 |
| 1 | 1 | 1 |
| 1 | 1 | 1 |

**(b)**

| | | |
|---|---|---|
| 1 | 2 | 1 |
| 2 | 4 | 2 |
| 1 | 2 | 1 |

**(c)**

| | | |
|---|---|---|
| 1 | 1 | 1 |
| 1 | 12 | 1 |
| 1 | 1 | 1 |

**Fig. 2** Masks of low-pass filters: mean filter (**a**), Gaussian filter (**b**) and other based on averaging filter (**c**)

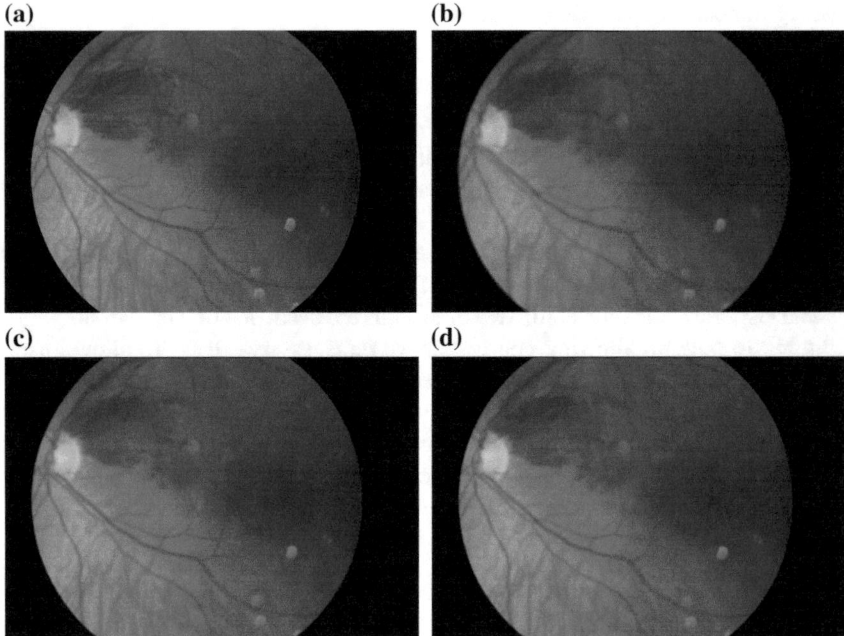


**Fig. 3** Different types of low-pass filters: original image (**a**), mean filter (**b**), Gaussian filter (**c**) and other based on averaging filter (**d**)

**Fig. 4** Mask of mean removal filter

| -1 | -1 | -1 |
|---|---|---|
| -1 | 9 | -1 |
| -1 | -1 | -1 |

- Gaussian filter uses values close to the discrete representation of the Gaussian function. The resulting blur is much higher than when using the averaging filter.

According to the name, high-pass filters enhance high-frequency components. The main purpose of using these filters is to emphasize the edges and details by increasing the contrast or increasing their brightness [7, 9]. The negative feature of this type of filter is the amplification of the noise present in the image. Visibility of the edges and details is also enhanced.

Example of this filter can be Mean Removal Filter. The use of this filter results in a sharpening of the edges. At the same time, this filter significantly strengthens the emerging noise (see Figs. 4 and 5). This is one of the most commonly used high-pass filters.

Statistical filters do not use the wave function. The current value is obtained by selecting the appropriate value under the mask [9]. Examples of such filters are:

- Median filter—the value of the current pixel is selected as the median value of all the pixels below mask. This filter is often used to derive because the resulting

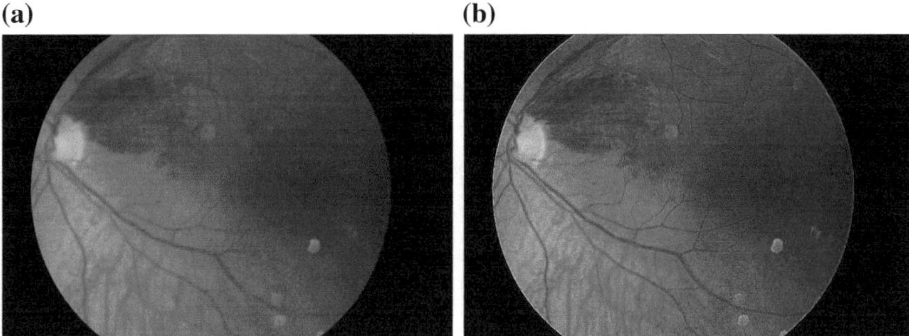

**Fig. 5** Original image (**a**) and image filtered using mean removal filter (**b**)

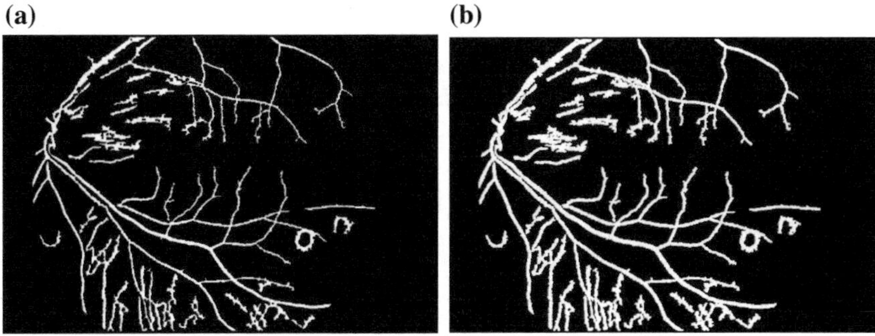

**Fig. 6** Erosion for black color: original image (**a**) and eroded image (**b**)

image is blurry to a lesser extent than using a high-pass filter. The effect of using this filter is to receive an image that resembles a painting painted with paint [8].

- Minimal filter—just like in a median filter, the value is selected among the pixels under mask. The difference lies in choosing the minimum value, not median.
- Maximum filter—value of the searched pixel is equal to the maximum value of the pixel below mask.

Another type of action in the image are the morphological operations that affect the form, the shape of the objects in the image. They are used when defining the shapes of objects in an image.

Examples of these operations are [7] the following:

- Erosion—when using the mask, if all the pixels under it have the same color, then test point also accepts the mask (Fig. 6). This operation is equivalent to using a minimum filter. Erosion reduces the field of the object to enlarge all kinds of noise inside the object.
- Dilation—If at least one pixel from the environment has the desired value, then the point that we consider also assumes that value (Fig. 7). A similar result can be reached using the maximum filter.

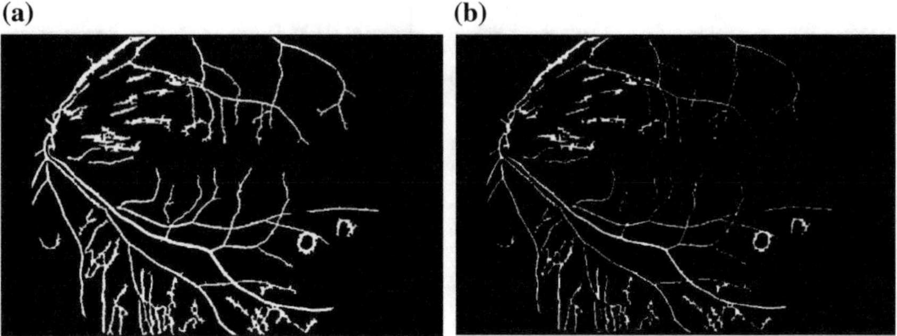


**Fig. 7** Dilatation for black color: original image (**a**) and eroded image (**b**)

- Opening and closing—these are the assemblies of two morphological functions, erosion and dilation in the correct order. These operations can be presented as follows:

$$\text{opening} = \text{erosion} + \text{dilatation} \tag{1}$$

$$\text{closing} = \text{dilatation} + \text{erosion} \tag{2}$$

The basis of many image processing algorithms is to change the color palette to grayscale. This is achieved by setting the value of three colors (channels) to identical values [5.13]. Choosing the right method may turn out to be crucial, so there are three methods to consider:

- By using one of the channels—depending on the channel selected, its value is replaced by the other two. The method is used in biometry, where one of the most used channels is the green channel.
- By using average channel values, the value is calculated by means of the mean. Theorem describes how to obtain the value set as each of the gray components:

$$Y = \frac{(R + G + B)}{3} \tag{3}$$

where

$R, G, B$ individual color frequencies of the channel data,
$Y$ value given as each of the components to obtain the gray color.

- With luminescence, the luminescence value is calculated using the formula (4). It reflects the way in which the human eye sees each of the components of light, respectively.

$$\mathrm{Y} = 0,299 * \mathrm{R} + 0,587 * \mathrm{G} + 0,114 * \mathrm{B} \tag{4}$$

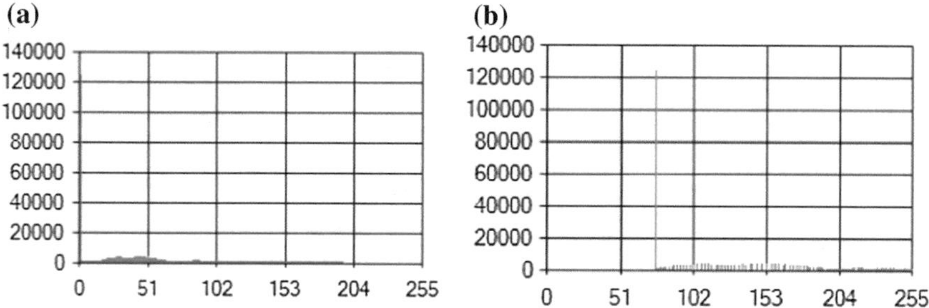


**Fig. 8** Histogram before (**a**) and after (**b**) equalization

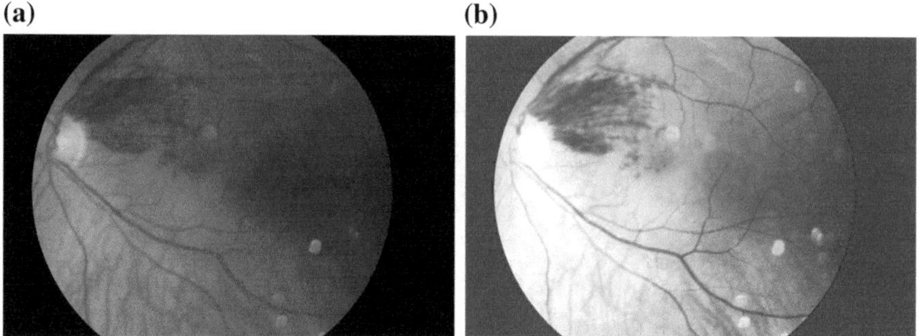


**Fig. 9** Equalization of histogram: image original (left) and with equalized histogram (right)

where

| | |
|---|---|
| *R*, *G*, *B* | values of individual colors in the examined point, |
| *Y* | value set as each of the components to get the gray color. |

Useful for research are histogram operations. Histogram is the graphical representation of numerical data distribution (Fig. 8). This is a bar graph showing the size of objects in class. In images, the histogram represents the number of pixels that contain consecutive values of the component color. Four histograms were created in the created program, three of which concerned the color distribution of red, green, and blue, and the values of previously discussed luminescence [8].

One of the operations performed using a histogram is equalization of histogram. This method is used to increase global contrast, especially useful when values represent narrowly focused data. It consists in spreading the range for a better readable image. This method does not recognize noise so that its number can increase significantly [7]. Despite the fact that often result of surgery is an image that contains unrealistic effects, it is useful in many aspects such as biomedicine or computer graphics [10] (Fig. 9).

Binarization is an operation aimed at converting a color image or a shade of gray to binary, e.g., black and white [3]. The most commonly done is by threshold, or

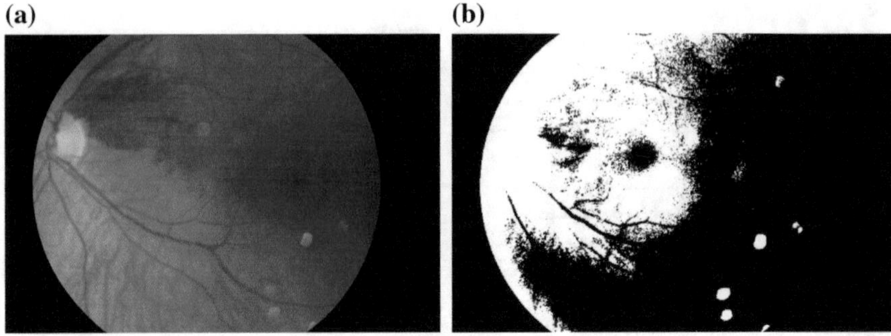


**Fig. 10** Otsu binarization: original image (**a**) and binarized (**b**)

classification is done by checking whether the value in a given pixel is less than or greater than the specified threshold.

$$g(i,j) = \begin{cases} 0\; for\; f(i,j) < T \\ 1\; for\; f(i,j) \geq T \end{cases} \tag{5}$$

where

- $i, j$ height and width of image,
- $g$ resulting image,
- $f$ start image,
- $T$ threshold value.

Another implemented method of binarization is the Otsu method. It focuses on selecting the threshold in such a way as to minimize variance in the two resulting classes [15]. All possible divisions are checked and the division with the lowest variances in classes [2, 7] is selected (Fig. 10).

## 3 The Authors' Algorithm

For research purposes, an application has been created to search for blood vessels in eyes retina. In order to determine the usefulness of a given sample, we have added algorithms for thinning and minutiae search. The retinal segmentation algorithm was based on the algorithm presented in [12], but it has a significant change—it does not use adaptive histogram compensation with contrast limitation. The general principle of the minutiae search algorithm is shown in the graph below (see Fig. 11).

The first step in the algorithm is to switch to shades of gray. Used for this purpose was the green image channel (see Fig. 12a). The image shows blood vessels to a degree that allows for further processing. The values of the individual component colors are replaced by green canal value.

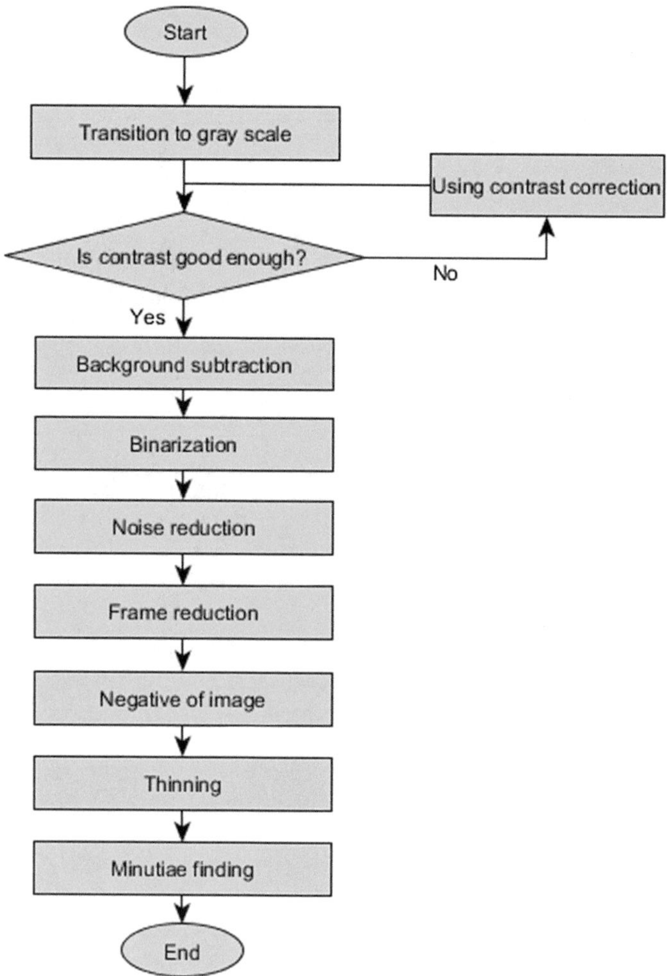


**Fig. 11** Algorithm diagram

The use of this component color is related to the fact that it is in the middle of the spectrum of visibility and that image in shades of gray using green color highlights the local contrast between the background and the foreground.

The next step is to improve the contrast in the image.

Instead, an optional step in this algorithm [12] is the mandatory step of applying adaptive histogram compensation with contrast limitation, but for research purposes this algorithm is not required to obtain satisfactory results.

If the image is poorly visible, because of the darkness of the image, the blood vessels can be used to improve the contrast. The LUT (Lookup Table) is used for

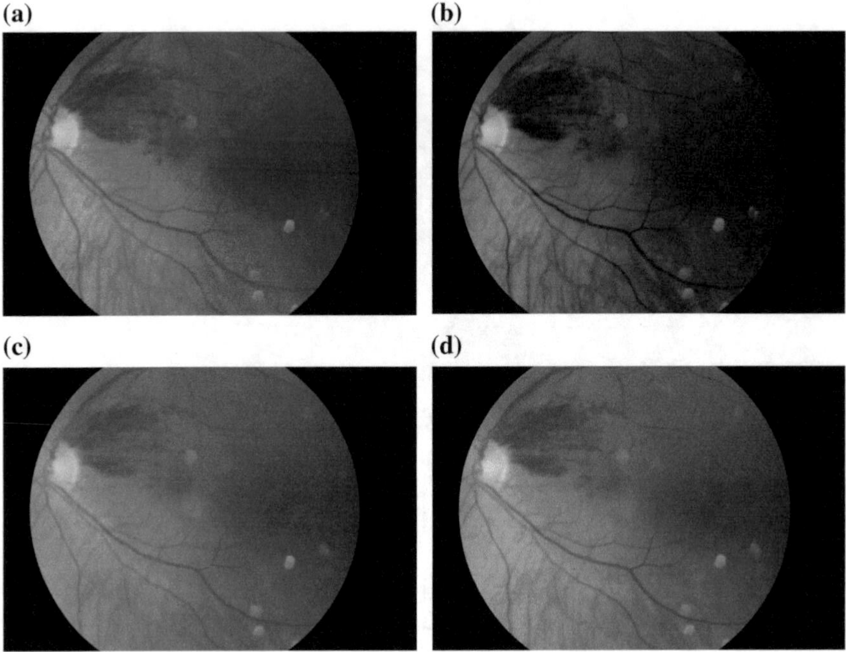


**Fig. 12** Image in grayscale: original image (**a**), using green color canal (**b**), using an average of channels value (**c**), using luminescence (**d**)

this, a discrete function that changes the values of the image components to the new values that are in the array. The function parameter can take three types of values.

The value used in the interval (0; 1) decreases the color value at all points, using parameter 1 will not make any changes, and the parameter greater than 1 will increase the contrast of the image. Use of the LUT table is as follows:

$$LUT(i)\begin{cases} 0, if\, a\left(i - \frac{i_{max}}{2}\right) + \frac{i_{max}}{2} < 0 \\ a\left(i - \frac{i_{max}}{2}\right) + \frac{i_{max}}{2}, if\, 0 \le a\left(i - \frac{i_{max}}{2}\right) + \frac{i_{max}}{2} \le i_{max} \\ i_{max}, if\, a\left(i - \frac{i_{max}}{2}\right) + \frac{i_{max}}{2} > i_{max} \end{cases} \tag{6}$$

where

$a$ parameter set by the user,
$i_{max}$ the maximum possible value of component color,
$i$ current value of the component color.

The next step in algorithm is to subtract the image from the initial image after the transformation with a low-pass filter with a width and height of 9 pixels. The middle pixel value is replaced by the average pixel value below the mask. The following equation is a mathematical representation of image averaging operations.

$$\text{h(x, y)} = \frac{1}{NxM} \sum_{i=1}^{NxM} gi \tag{7}$$

where

$x, y$ height and width of the image,
$M, N$ height and width of the mask,
$i$ number of pixel under the mask,
$g$ color value of the pixel under the mask,
$h$ the resulting image.

The resulting image only stores values in which the original image was higher than the image obtained with the scavenging filter. The final image is obtained by the formula (8).

$$k(x, y) = \begin{cases} h(x, y) - g(x, y) if h(x, y) - g(x, y) > 0 \\ 0 \, if \, h(x, y) - g(x, y) \leq 0 \end{cases} \tag{8}$$

where

x, y width and height of the image,
k the resulting image,
h the initial image,
g the fuzzy image.

Then use the binary algorithm. The implemented algorithm for binarization is the Otsu method. Its function is to choose the threshold so that in both classes received variance was as small as possible. When performing the algorithm, all divisions are checked and the best one is selected (see Fig. 14b).

Much of the algorithms used to emphasize the contrast also increase the noise in the image. After binarization in the image, blood vessels, a frame around the image and considerable noise are visible. In order to remove the areas in the image not related to the blood vessels, an algorithm was created to remove areas that are larger than the threshold (see Fig. 14c).

First, the algorithm passes each pixel to verify that it is not part of the background, if it is not checked whether it has a numbered neighbor, if it has a field that accepts the lower of these values. If there are no neighbors who have a given number, then he himself receives a unique new number.

The result is an array where all areas have numbers, but some have more class numbers than one, so the next step is to create a neighborhood list and join neighboring classes by changing the values of the larger ones to the smaller ones. After this step, all areas are described with one unique class area. After summing up the number of class instances, those that are smaller than the threshold are removed. Depending on the image and its quality, different thresholds were used, in this work the threshold was 100.

**Fig. 13** Sum counting mask for minutiae research

| 1 | 2 | 4 |
|---|---|---|
| 128 | 256 | 8 |
| 64 | 32 | 16 |

In the analysis of the vessels disturbance still remains after previous operations frame. It was removed by binarization of the original image at a low threshold setting and the use of erosion in this image to enlarge the area to be removed. After surgery, only blood vessels remain, which may be frayed. A morphological filter called closure [3] was used to merge areas that were not removed.

For better visibility, the next step in the algorithm is to make a negative image, which limits the color of the image from white to black and vice versa (see Fig. 14d). This allows for better visibility of subsequent changes in the image. The next phase of preparation of the images for the study is blood vessel thinning. For this purpose, the K3M algorithm was implemented. It uses the sum of the neighbors of the examined pixels to determine if a given pixel should be removed or not. The result of this algorithm is the image of vessels in form of a skeleton, e.g. in the form of a line of one-pixel thickness [13] (see Fig. 14e).

The final step in the algorithm is finding a minimum. This is done in a similar way to thinning. A mask (see Fig. 7) is used to easily determine which pixels are underneath it. A sum determines the specific point system. Then it is checked whether the sum is in one of the three letters responsible for storing the sums classifying into the respective classes (Fig. 13).

If the sum is in the list the point along with the closest neighbors are colored to the right color. The algorithm searches for three types of minima: intersection, branching, and end/start. In the picture (see Fig. 14f), the intersection is shown in yellow, blue start and end, and red fork. The code below shows the values at which the corresponding minutiae type is assigned to the pixels.

```
int[] R = {293, 329, 338, 404, 277, 340, 337, 325, 420,
297, 330, 402};
int[] KP = {288, 258, 384, 264, 257, 320, 272, 260};
int[] S = {170, 341};
```

where:

R fork,
KP start or end,
S intersection.

## 4 Research Results

During the study, the samples were divided into 4 groups, depending on the disease occurring. Each group is considered independently. Each sample was examined for

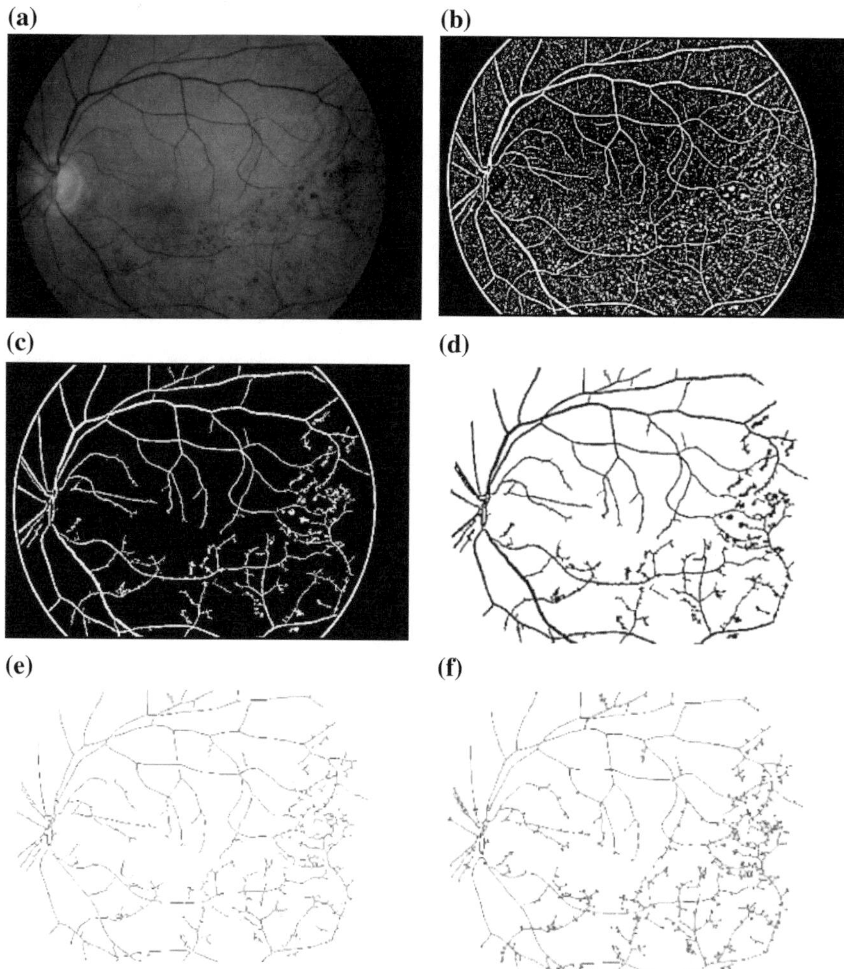


**Fig. 14** Steps of the algorithm: grayscale using green canal value (**a**), background subtraction (**b**), image after noise removal (**c**), negative of the image after frame removal (**d**), image after thinning (**e**), blood vessels with minutiae found (**f**)

the distance from the center of the image and the distance from the edge of the image in which the blind spot is located—the point at which the optic nerve leaves the eyeball. It is also the place where the largest accumulation of large blood vessels in the eye is located. Each image will show the results of the search operation as well as the number of minutiae found. The amount of minimum used is the amount that occurs in a given area. Two methods of minutiae qualification have been selected for the respective group (see Fig. 15).

The first method makes the minutiae class dependent on its Euclidean distance from the center of image. If minutiae are in the center of image, errors related to the

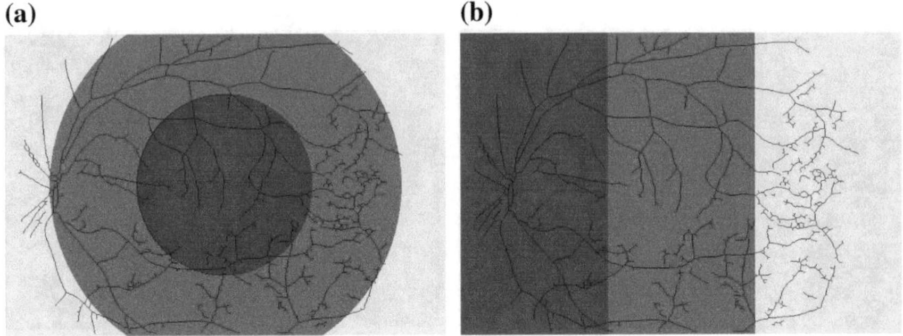


**Fig. 15** Visualization of minutiae classification: distance from center of image (**a**), distance from blind spot (**b**)

mapping of points from the inside of the sphere in the plane are minimal. The image is farther from the center of the image, the bigger the error, and the results become less accurate.

The second method uses the division of the image into three sectors by dividing the width into three equal sections. The method allows you to assign a minutia point to the appropriate class depending on the distance from the blind spot.

Results for images showing arterial thrombosis in the retina region demonstrate a low incidence of the disease in detecting blood vessels and minutiae. Pictures show the following typical changes during complete closure of the blood vessel, e.g., retinal bleeding at the site of ischemia. The dish in a place where no blood flows itself becomes brighter and harder to detect, or even undetectable. In some cases, it is not possible to diagnose a disease without contrast imaging. The images obtained in this manner are in shades of gray, with the vessels behind the embolism site darker than the rest.

Some samples show the retina from both eyes of one person so it is possible to notice that lesions usually affect only one eye, so it is likely to recognize a person using a healthy eye. During the study also appeared noise, some of which are associated with selected or other disease units. In the case of embolism, the most common noise comes from discoloration of the retina and reflexes resulting in the faulty qualification of the object as a blood vessel. However, these are noises that can easily be removed. They have a distinctly different color from the environment.

The sum in each biometric sample tested exceeds 150 minutiae. In most countries, twelve minutiae are enough to correct a person, in some cases even less, if they are rare.

In addition, cavities in blood vessels appear, which are caused by a change in the color of the vessel to the lighter. In most cases, an embolism is formed in the artery of a blood vessel, which limits its extent to one blood vessel.

Clots, like embolism, are usually located in one eye of a sick person. The disease is characterized by changes in color, slightly darker than blood vessels called droplets. Other than the dullness in the picture, there may be other reasons for noise and falsification. As with embolism, there are also light reflections and discoloration of the retina, not associated with blood clots. In research samples, a perception of petechiae in various places of the retina. These changes are concentrated in the area of one blood vessel, but it is also possible to change the color of the retina to that of the blood vessels in a larger area.

In most cases, it is possible to search for more blood vessels. The problem may be diarrhea that arises both in the area of the blood vessels as well as outside, so that the attempt to remove the noise may end up removing the blood vessels from the picture. High-grade blood clots can cause significant difficulties in the search for blood vessels to the extent that it is impossible to find one minutiae. Depending on how to remove the noise caused by the disease, it is possible to recognize a person using a sample from the patient's eye. Depending on the location of the lesions around the blood vessels, the decomposition may take different forms.

A small number of specimens in the case of spouts did not detect the trend on a larger scale, but it can be concluded that noise in the image may be associated with the area where the blood is located. In particular the border of the place of hemorrhages. In the case of samples obtained, it is noticeable that the place where the nozzle is located is not considered to be the vessel. The veins and arteries are negligibly shaded by the change. The image does not have any other defects beyond the previously discussed. Dishes have continuity, but imperceptible are the vessels behind the spout.

The largest group of all samples are those with retinal detachment. Depending on the type and place of occurrence, the results may vary. There are cases in which finding the vessels is difficult because of the high number of noise, covering the vessels by changing the disease as well as blurring the remaining blood vessels, which results may prove to be unreliable.

In other cases, it is possible to obtain satisfactory results. The area of change is dependent on the severity of the disease, so in each case, the results should be treated individually.

In the case of decolorization, a new type of noise resulting from the lesion itself appears, which may be caused by a border of lesions, discolored lesions on lesion, or by transient changes in the lesion that may have been different from the lesion.

Due to the presence of noise or the inability to reach the required number of minutiae, it may not be possible to confirm the identity of the sick person.

Places of occurrence of the minutiae depend very much on the location of occurrence of deodorization and its size.

## 5 Summary and Conclusions

The aim of this study was to examine the influence of diseases on the samples, e.g., the images of the fundus. This task was accomplished by creating an application that enables numerous imaging operations as well as through research. During the study, a thorough review of the possible algorithms used to look for blood vessels in the retina image was made. Also, attention was paid to numerous image processing operations and their results. By adding to one of the algorithms of such elements as thinning and minutiae search possible further analysis related to the quality of blood vessels. The application allows for further functional development, such as identity verification and disease diagnosis.

Diseases described in this article represent a small part of the total number of diseases that may occur. Their choice was not accidental. These pathologies are one of the most common. In addition, it is worth noting that some of them are not curable and no reason for their formation is known. Diseases such as thrombosis and blood clots are associated with specific blood vessels, so their appearance is usually confined to one eye, but this is not the rule.

Changes made during detachment can greatly influence the view of vessels in the area in which they occur. In some cases, vessels throughout the retina are not able to be found using an algorithm written for the purpose of this work. Changes caused by the disease can also affect the vessels outside their area of occurrence. The veins and arteries may become less visible, which can result in the accumulation of far too much minutiae in one area.

The results, however, appear in the presence of sputum. Blood vesicles cover the vessels behind it so it is impossible to find the right minutiae in this area.

All these diseases tend to start out within one eye. As soon as the disease develops, there is a possibility of change in the second. Each of these changes is unitary and the usefulness of the patient's retinal biometric image should be made individually.

After consulting the results with the ophthalmologists, it can be stated that the results of the algorithm are satisfactory. All elements work according to their purpose and the final result allows to draw useful biometric conclusions.

**Acknowledgements** This work was supported by grant S/WI/3/2018 from the Białystok University of Technology and funded with resources for research by the Ministry of Science and Higher Education in Poland.

## References

1. Agarwal, A.: Gass' Atlas of Macular Diseases One. Elsevier, USA (2012)
2. BahadarKhan, K., Khaliq, A.A., Shahid, M.: A morphological hessian based approach for retinal blood vessels segmentation and denoising using region based otsu thresholding. Plos one **11**(7) (2016). Article Number: e0158996
3. Boulgouris, N.V., Plataniotis, K.N., Micheli-Tzanakou, E.: Biometrics. Theory, Methods, and Applications. Institute of Electrical and Electronics Engineers, Inc (2010)
4. Daniel, E., Ying, G., Siatkowski, R.M., et al.: Intraocular hemorrhages and retinopathy of prematurity in the telemedicine approaches to evaluating acute-phase retinopathy of prematurity (e-ROP) study. Ophthalmology **124**(3), 374–381 (2017)
5. Faizalkhan, Z., Nalini Priya, G.: Automatic segmentation of retinal blood vessels employing textural fuzzy C-means clustering. In: IEEE International Conference on Emerging Technological Trends (2016)
6. Gill, R., Kaur, I.: Segmentation of retinal area by adaptive SLIC superpixel. In: 1st IEEE International Conference on Power Electronics, Intelligent Control and Energy Systems (2016)

7. Gonzalez, C.R., Woods, E.R.: Digital Image Processing. Pearson Education (2009)
8. Gupta, N., Asarti, E.: Performance evaluation of retinal vessels segmentation using a combination of filters. In: 2016 2nd International Conference on Next Generation Computing Technologies (NGCT-2016) (2016)
9. Jansonius, N.M., Cervantes, J., Reddikumar, M., Cense, B.: Influence of coherence length, signal-to-noise ratio, log transform, and low-pass filtering on layer thickness assessment with OCT in the retina. Biomed. Opt. Express **7**(11), 4490–4500 (2016)
10. Nishima, Gill, R., Kaur, I.: Segmentation of retunal area by adaptive SLIC superpixel. In: IEEE International Conference on Power Electronics, Inteligeent Control on Energy System (ICPEICES-2016)
11. Priya, R., Priyadarshini, N., Rajkumar, E.R., Rajamaniy, K: Retinal vessel segmentation under pathological conditions using supervised machine learning. In: Proceedings of 2016 International Conference on Systems in Medicine and Biology
12. Saleh, M.D., Eswaran, C., Mueen, A.: An automated blood vessel segmentation algorithm using histogram equalization and automatic threshold selection. PMC (2011)
13. Tabędzki, M., Saeed, K., Szczepański, A.: A modified K3M thinning algorithm. Int. J. Appl. Math. Comput. Sci. AMC **26**(2), 439–450 (2016)
14. Torres Seriano, M.E., Garcia Aguirre, G., Gordon, M., Kon Graversen, V.: Diagnostic Atlas of Retinal Diseases. Bentham eBooks imprint (2017)
15. Wachulec, P., Saeed, E., Bartocha, A., Saeed, K.: Retinal feature extraction with influence of its diseases on the results. Applied Computation and Security Systems Volume One. Advances in Intelligent Systems and Computing. Springer, pp. 37–48

# Part II
# Signal Processing and Analytics—I

# Analysis of Stimuli Discrimination in Indian Patients with Chronic Schizophrenia

**Jaskirat Singh, Sukhwinder Singh, Savita Gupta and Bir Singh Chavan**

**Abstract** Event-Related Potentials (ERPs) components are widely studied to understand brain response associated with cognitive processes in terms of amplitude, latency, and brain topography. Schizophrenia is a psychiatric illness which is characterized by reduced amplitude and increased latency of P300 ERP component. In this work, we have analyzed the attributes of P300 component involved during discrimination of visual stimuli (frequent, infrequent, and rare) at midline frontal and parietal sites in 14 schizophrenic patients who were receiving remediation in psychiatric setup for a long time. The key findings include increase in P300 amplitude when infrequent and rare stimuli were presented which reflected better contribution of attentional resources toward discrimination of stimuli. The P300 latency was found to be increased at same measurement sites for standards and rares for which the responses have to be ignored. The latencies were reduced in the case of infrequent stimuli (targets) which reflected better speed of cognitive processes involved during the identification of the stimuli.

**Keywords** Schizophrenia · Event-Related Potentials · P300 · Amplitude Latency

J. Singh · S. Singh (✉) · S. Gupta
University Institute of Engineering and Technology, Panjab University, Sector 25, Chandigarh, India
e-mail: sukhdalip@pu.ac.in

J. Singh
e-mail: jaskiratj@gmail.com

S. Gupta
e-mail: savita2k8@yahoo.com

B. S. Chavan
Department of Psychiatry, Government Medical College and Hospital, Sector 32, Chandigarh, India
e-mail: drchavanbs@gmail.com



R. Chaki et al. (eds.), *Advanced Computing and Systems for Security*, Advances in Intelligent Systems and Computing 897,
https://doi.org/10.1007/978-981-13-3250-0_4

## 1 Introduction

Electroencephalography (EEG) is a noninvasive method to measure electrical activity produced by the brain. When a person is exposed to a stimulus either visual or auditory, or sensory, the response is elicited from the brain which is measured as changes in the parameters of the electric potential over the scalp. Event-Related Potentials (ERPs) are tiny fluctuations (millisecond resolution) that are burred within the ongoing EEG which are later separated by the signal averaging method, exposing +/− peaks that constitute the cortical response to the event of interest. ERP components (P50, N100, N200, P300, N400 and Mismatch Negativity) which are strictly time and phase locked to the onset of particular stimulus are elicited at different stages of information processing: early (<100 ms), pre-attentive (200 ms), and late stages (>300 ms). The simplest ERP parameters are latency (how long after the event they appear), direction (positive or negative), amplitude (the strength of the voltage change), and topological distribution of the component over the head surface. An abnormality in the amplitude, latency, and phase synchrony of these components quantifies the abnormal brain activity and functional disconnectivity [1]. Out of all the components, P300 is the positive electric potential recorded at the late stage of the testing period (the peak occurs about 300 ms after the stimuli presentation) when the subject identifies a target stimulus from among a large number of nontarget stimuli. For P300 to occur, the subject must attentively monitor the target stimuli and react to each occurrence of the stimulus by pressing a key.

Schizophrenia is a psychiatric illness which is characterized by the breakdown of thought process. P300 component has been extensively studied as an important neurophysiological feature of schizophrenia because this component reflects attention control and memory, both required for final evaluation of stimulus are found to be abnormal in schizophrenia [2, 3]. Ford et al. (1999) showed P300 amplitude is related to clinical state with severely ill patients having smaller P300s than moderately ill patients and scored higher on Brief Psychiatric Rating Scale (BPRS) scores [4]. Rao et al. (1995) found smaller auditory P300 amplitude in schizophrenic patients who were in stable clinical remission state at both central and parietal recording positions [5]. Turetsky et al. (2000) compared ERPs from patients to those from their own healthy siblings. Patients had reduced parietal and frontal auditory P300 amplitudes. The healthy siblings of the schizophrenic probands showed reduction in frontal P300 only. Prolongation of P300 latency has also been reported in schizophrenia [6]. P300 topographic asymmetries were reported in unmedicated schizophrenics [7]. Current status on P300 reduction in schizophrenia is in line with the studies in the past. Sahoo et al. (2016) studied auditory P300 ERP component in Indian patients. The authors showed reduced amplitude and increased latency at Cz and Pz sites distinguished patients with schizophrenia in remission from patients with Acute and Transient Psychiatric Disorder (ATPD) and healthy controls [8]. Alexis McCathern (2017) showed impaired P3b component during selective attention to absent stimuli on a tone counting task in both long-term schizophrenia and first episode as compared to their matched control groups [9]. Kim et al. (2017) reported P300 amplitude is an endophenotype for schizophrenia and greater intertrial variability of P300 was associated with more severe negative and cognitive symptoms in schizophrenia patients. Altered P300 topographic symmetry has also been seen in schizophrenia [10]. Reduction in P300 amplitude is also reported to lateralize over the left hemisphere temporal region as compared to right temporal region [11]. The P300 component reflects the process in which attention-driven memory comparison is derived to find out variations in the features of incoming stimuli. These alterations related to the changes in the stimuli are reflected in the background EEGs/ERPs [12].

There are only very limited studies in the literature that have analyzed event-related potentials in the Indian sample. Auditory odd-ball paradigm has already been studied for the same [5, 8]. This study focuses on the analysis of visual odd-ball paradigm to record alterations in P300 component elicited during classification of different stimulis: frequent, infrequent, and rare.

## 2 Materials and Methods

**(a) Subjects**: Fourteen patients (eight female and six male: Mean Age: 36.4 years) diagnosed with Schizophrenia in accordance to International Classification of Mental and Behavioral Disorder (ICD-10) were recruited from neurocognitive rehabilitation facility—Disability Assessment and Rehabilitation Triage (DART), Department of Psychiatry, Government Medical College and Hospital (GMCH), Sector 32, Chandigarh, India. The inclusion and exclusion criteria for the study are stated as follows:

**(1) Inclusion criteria**:

(i) Age: 18–60 years.
(ii) Duration of illness: minimum 2 years

**(2) Exclusion criteria**:

(i) Patients with comorbid neurological and physical illness.
(ii) Substance abuse except for nicotine dependence
(iii) Patients with head injury

All the subjects gave informed consent to participate in the study. The study was approved by the local research and ethics committee. All the subjects were on regular medication regime: typical/atypical antipsychotics. Also, the subjects were receiving remediation in the psychiatric setup for a long time (more than a year).

**(b) Experimental Task**: Figure 1 shows the ERP experiment selected for the study has been adapted from [13, 14]. This is a classical P300 odd-ball experiment that emphasizes on visual ERPs. In this task, letters "**O**" and "**Q**" are displayed to a subject one at a time. The "**O**" letter (standards/frequent) appears more frequently 80% of the time than the letter "**Q**" (Infrequent) which appears 20% of the time. Also, letter "⊖" (Rare) is used as distractor, that share the same occurring phenomenon as targets stimuli unexpected but have lower occurring probability than the targets

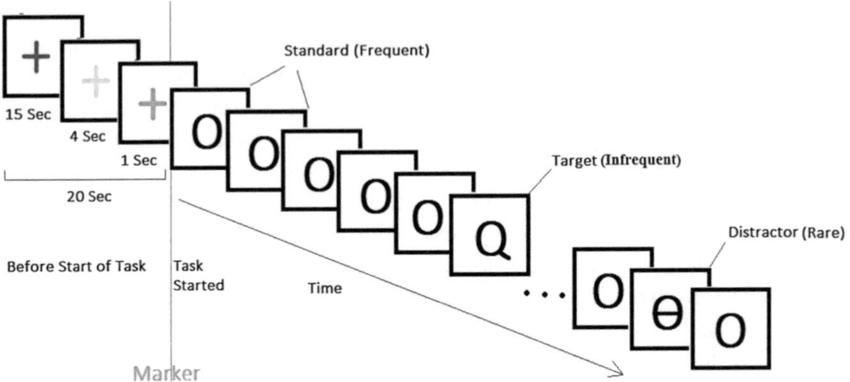


**Fig. 1** Experimental visual odd-ball task

stimuli. Subjects were instructed to respond to targets only by touching the button displayed on the screen as soon as possible. The task was presented on a 7 inch tablet as shown in Fig. 2, Android OS and stimuli presentation program was specifically written for the same. For each stimuli, the type and timestamps where recorded with respect to the task startup marker. Each stimuli had 500 ms duration with a random interstimulus interval of 1000 ms. In order to make the subjects aware about the beginning of the task, the first task stimuli was preceded by a cross sign whose color changed from red (after 15 s) to yellow (4 s) and later to green (1 s) for a total span of 20 s. Subjects were instructed to maintain fixation of eyes on the cross sign and the task will begin as soon as the cross sign turns green. All the instructions were given to the subjects in the language comprehended by them (Hindi or Punjabi).

**(c) Methodology**: The subjects were comprehensively explained about EEG recording and all their queries regarding the procedure were answered. All the subjects were instructed to come with clean dry hair and not to apply oil on the day of test. The subjects sat on an armchair and were told to limit blinking of eyes. The electrical activity was recorded from 19 Ag–Cl electrodes (FP1, Fp2, F7, F3, Fz, F4, F8, T3, C3, Cz, C4, T4, T5, P3, Pz, P4, T6, O1, and O2) placed over the scalp and ECG was additionally recorded using Clarity 10–20 EEG recording system. The EEG was digitized at sampling frequency of 256 Hz. All the measurements were taken following CZ referential montage as recommended by [14]. The ground electrode was placed on the Fpz location (forehead). During the experiment, the impedance of the electrodes lied between 6 and 20 k. Bad electrodes were interpolated with average from other electrodes. The continuous EEG data was band passed at 0.01–100 Hz and notch filtered at 50 Hz during recording. The data was subjected to Independent Component Analysis (ICA) for detailed signal component analysis. From all the existing ICA algorithms, AMICA and InfoMax Algorithms were shown to generate precise component topographies with greater mean mutual information reduction rate [15]. In this work, the InfoMax (runica) method has been applied for ICA

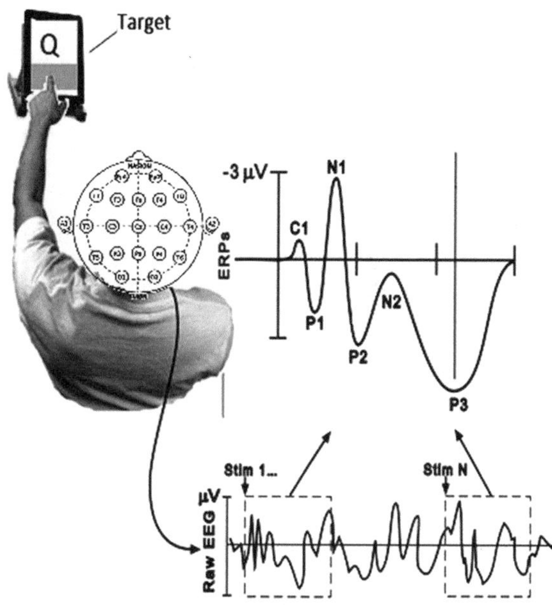


**Fig. 2** Electrophysiological 10–20 recording

decomposition. The artifact components related to eye blinks, lateral eye movement, muscle activity and cardio artifacts were rejected by visual inspection of component topographies and their respective power spectrum. ICA components corresponding to Ocular activity (low frequency, high amplitude signal observed in Frontal Fp1 and Fp2 electrode sites) were removed after inspecting their power spectrum. Muscle artifacts having low amplitude and wide frequency range (mainly observed in temporal and parietal regions [16]) were also removed. ADJUST and SemiAutomated Selection of Independent Component Analysis (SASICA) were also utilized for assistance in identification and rejection of artifact-related components [17, 18]. The EEG data was segmented and categorized (standard, target, and distractor) into stimulus-locked epochs of duration 1 s each (−200 to 800 ms baseline corrected for pre-stimulus recording time) as per the recorded event information. The epochs were rejected that have electrical potential greater than +/− 75 V. The average number of trials left after artifact rejection is 124 out of total 160. The EEG data was processed in EEGLAB, a MATLAB-based open-source program for signal analysis [19].

## 3 Results and Discussion

**(a) Grand Average**: Epochs averaging is a standard procedure to estimate ERPs as it reduces trial to trial variability and background noise. The epochs were averaged offline using ERPLAB plug-in for EEGLAB [20]. The ERPs were computed separately and binning is done for each type of stimuli: standard, target and distractor.

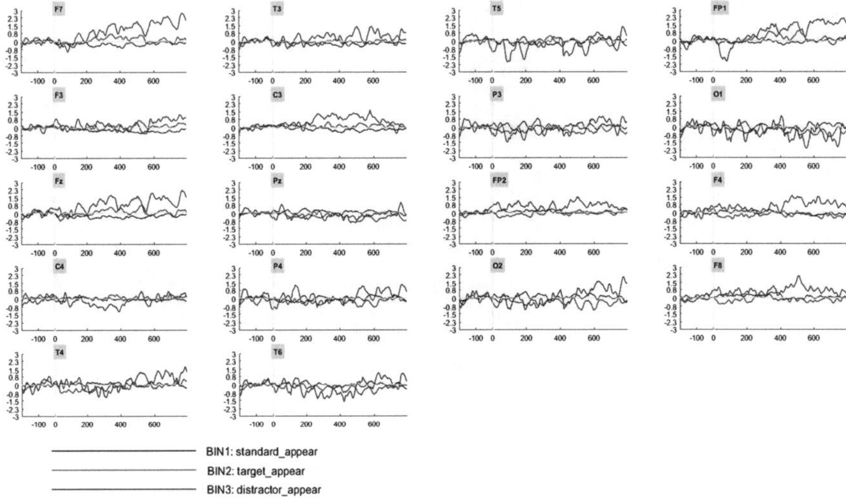


**Fig. 3** Grand averages calculated across ERPsets

From multiple, ERPsets (across subjects), each bin type was averaged individually with the similar bin from other ERPsets resulting into a grand averaged ERPset as shown in Fig. 3. This grand average (group ERPs) represents the average of individual ERPs of all the subjects in the study obtained over all the electrode sites. It is from this grand average the peak amplitude, latency, and topographic scalp maps are further computed to observe ERP components in the schizophrenic patient group as whole when different stimulis (standard, target, and distractor) was presented. For this experiment, P300 features were measured at the midline Frontal (Fz) and Parietal (Pz) regions at which they are considered to be maximal [2, 21].

A topographic map is a representation in which spatial location of activity within a neuronal system is highlighted when parallel processing of stimuli by different regions of brain is taking place. Therefore it useful to observe the grand average (group) topographical maps within the measurement window of 250,420 ms. Figure 4 displays the distribution of potential over the scalp regions under frequent (standard), infrequent (target) and rare (distractor) conditions at different temporal scales: 250–300 ms, 300–350 ms, 350–400 ms, and 400–420 ms. With the progression of time window (50 ms), the shift in distribution of electric potentials can be observed from posterior regions to anterior regions of the brain in case of frequent stimuli. When the targets appear, the potentials get widely distributed over left posterior central parietal regions and in case of distractor the potentials can be seen to be localized over left anterior frontal electrode sites which indicate increase in cognitive demands as more concentration is required during the identification of the stimuli.

**(b) Amplitude and Latency**: The P300 component was calculated as the largest positive deflection occurring within the time window between 250 and 420 ms [7]. The peak amplitude was measured relative to the pre-stimulus baseline. The peak latency was defined as the time from stimulus onset to the peak of the P300

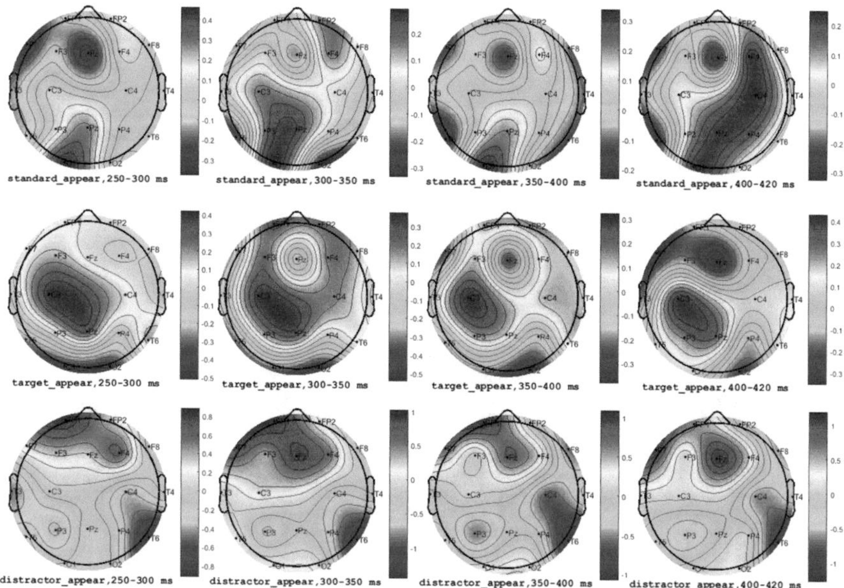


**Fig. 4** Scalp topological plots for standard, target, and distractor stimuli (window 250–420 ms, step size of 50 ms)

component. After presentation of each stimulis, there are major differences between the peak amplitude measurements at midline Frontal electrode site (Fz) as shown in (Fig. 5a, b). The mean peak amplitude in the measurement window of 250–420 ms was 0.378 V for standards, 0.907 V for targets, and 2.538 V for distractor. Similar increase in amplitude is observed for the rare and infrequent stimuli at Pz electrode site (Fig. 5c, d). The increase in amplitudes for the targets (infrequent) and distractors (rare) in comparison to the frequent stimulus is indicative of better allocation of cognitive resources during identification of stimuli.

The latency in the same measurement windows at Fz is observed to be higher in case of distractor (360 ms), standards (338 ms) and relatively lower in targets (326 ms) (Fig. 6a, b). Since, both Target and Distractor share the same characteristics-being rare and infrequent, they are expected to share the similar outcome. But, in our case the latency in target is lower and in distractor it is observed to be higher. Similarly from Fig. 6c, d, the latencies at Pz sites are also observed to be higher for rare and frequent stimulus: distractor (344 ms) and standard (352 ms). Since, the distractor occur with lower frequency than the targets, the time taken by the cognitive processes to classify the stimuli as target is less as compared to the distractor (requiring more cognitive demands/resources). The increased latency is understood to be due to involvement of inhibitory processes during selection of response after the stimulus has been presented. Since, the subjects were mainly asked to make a response when the targets appear, the observed increased amplitude and reduced latency in case of target stimuli correspond to signs of improvement in selective attention and speed of processing in patients with schizophrenia.

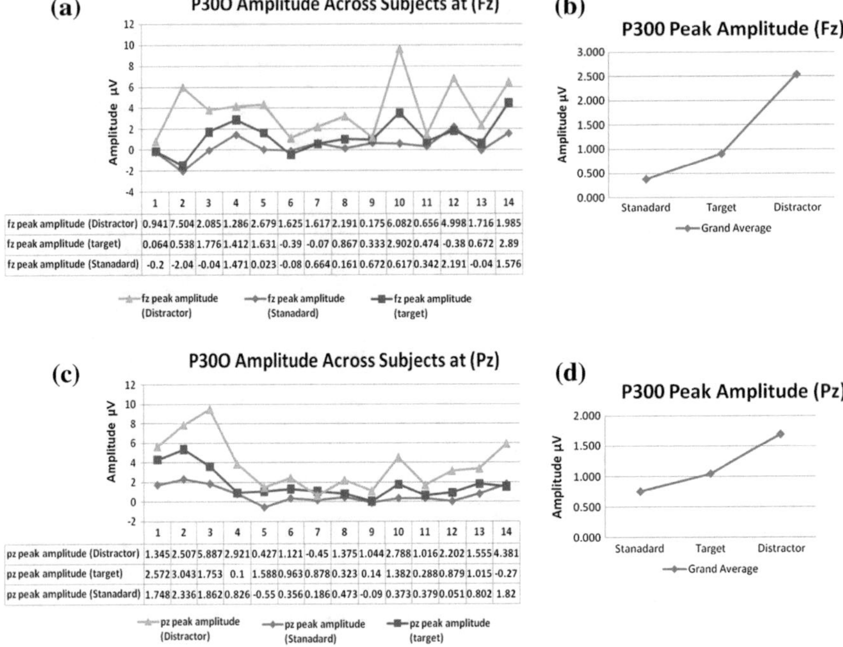


| | 1 | 2 | 3 | 4 | 5 | 6 | 7 | 8 | 9 | 10 | 11 | 12 | 13 | 14 |
|---|---|---|---|---|---|---|---|---|---|---|---|---|---|---|
| fz peak amplitude (Distractor) | 0.941 | 7.504 | 2.085 | 1.286 | 2.679 | 1.625 | 1.617 | 2.191 | 0.175 | 6.082 | 0.656 | 4.998 | 1.716 | 1.985 |
| fz peak amplitude (target) | 0.064 | 0.538 | 1.776 | 1.412 | 1.631 | -0.39 | -0.07 | 0.867 | 0.333 | 2.902 | 0.474 | -0.38 | 0.672 | 2.89 |
| fz peak amplitude (Stanadard) | -0.2 | -2.04 | -0.04 | 1.471 | 0.023 | -0.08 | 0.664 | 0.161 | 0.672 | 0.617 | 0.342 | 2.191 | -0.04 | 1.576 |

| | 1 | 2 | 3 | 4 | 5 | 6 | 7 | 8 | 9 | 10 | 11 | 12 | 13 | 14 |
|---|---|---|---|---|---|---|---|---|---|---|---|---|---|---|
| pz peak amplitude (Distractor) | 1.345 | 2.507 | 5.887 | 2.921 | 0.427 | 1.121 | -0.45 | 1.375 | 1.044 | 2.788 | 1.016 | 2.202 | 1.555 | 4.381 |
| pz peak amplitude (target) | 2.572 | 3.043 | 1.753 | 0.1 | 1.588 | 0.963 | 0.878 | 0.323 | 0.14 | 1.382 | 0.288 | 0.879 | 1.015 | -0.27 |
| pz peak amplitude (Stanadard) | 1.748 | 2.336 | 1.862 | 0.826 | -0.55 | 0.356 | 0.186 | 0.473 | -0.09 | 0.373 | 0.379 | 0.051 | 0.802 | 1.82 |

**Fig. 5** P300 Amplitude across subjects at Frontal (Fz) and Parietal (Pz) electrode sites

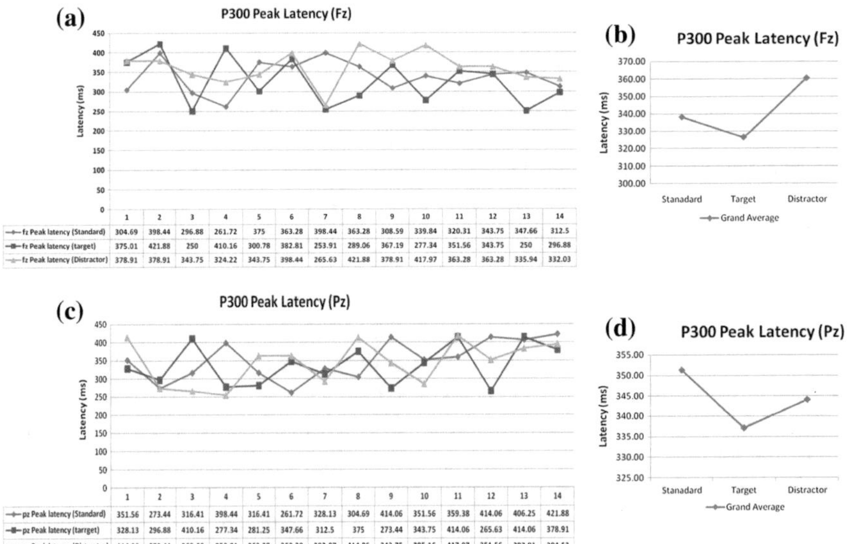


| | 1 | 2 | 3 | 4 | 5 | 6 | 7 | 8 | 9 | 10 | 11 | 12 | 13 | 14 |
|---|---|---|---|---|---|---|---|---|---|---|---|---|---|---|
| fz Peak latency (Standard) | 304.69 | 398.44 | 296.88 | 261.72 | 375 | 363.28 | 398.44 | 363.28 | 308.59 | 339.84 | 320.31 | 343.75 | 347.66 | 312.5 |
| fz Peak latency (target) | 375.01 | 421.88 | 250 | 410.16 | 300.78 | 382.81 | 253.91 | 289.06 | 367.19 | 277.34 | 351.56 | 343.75 | 250 | 296.88 |
| fz Peak latency (Distractor) | 378.91 | 378.91 | 343.75 | 324.22 | 343.75 | 398.44 | 265.63 | 421.88 | 378.91 | 417.97 | 363.28 | 363.28 | 335.94 | 332.03 |

| | 1 | 2 | 3 | 4 | 5 | 6 | 7 | 8 | 9 | 10 | 11 | 12 | 13 | 14 |
|---|---|---|---|---|---|---|---|---|---|---|---|---|---|---|
| pz Peak latency (Standard) | 351.56 | 273.44 | 316.41 | 398.44 | 316.41 | 261.72 | 328.13 | 304.69 | 414.06 | 351.56 | 359.38 | 414.06 | 406.25 | 421.88 |
| pz Peak latency (tarrget) | 328.13 | 296.88 | 410.16 | 277.34 | 281.25 | 347.66 | 312.5 | 375 | 273.44 | 343.75 | 414.06 | 265.63 | 414.06 | 378.91 |
| pz Peak latency (Distractor) | 414.06 | 273.44 | 265.63 | 253.91 | 363.28 | 363.28 | 292.97 | 414.06 | 343.75 | 285.16 | 417.97 | 351.56 | 382.81 | 394.53 |

**Fig. 6** P300 Latency across subjects at Frontal (Fz) and Parietal (Pz) electrode sites

## 4 Conclusion

In this study, three stimulus odd-ball tasks on patients with schizophrenia were administered and visual ERPs were extracted from the raw EEG. The P300 is an ERP component that is noticeable when attention is engaged toward the changes in the environment/stimuli. The abnormalities in P300 ERP components, i.e., reduced amplitude and increased latency has been widely reported, though very limited studies are available that replicates the signature P300 abnormalities in Indian schizophrenic patients. The available studies have recognized P300 abnormalities in an auditory odd-ball paradigm. In this article, P300 peak amplitude and latency under different stimuli conditions (frequent, infrequent and rare) are computed in the measurement window of (250–420 ms). The principal finding from this study is that ERP components change in peak amplitude, peak latency and topological distribution across the different stimuli types. Compared to the frequent stimuli, the amplitudes are found to be higher in case of infrequent and rare stimulis at frontal and parietal regions. On the other hand, latencies were found to be decreased for the infrequent stimuli (targets). The increase in P300 amplitude reflects contribution of attentional processes in the discrimination of stimuli. The reduction in latencies is also observed during the identification of target stimuli. Also, topological distribution of potential over the scalp regions under frequent (standard), infrequent (targets) and rare at different temporal scales: 250–300 ms, 300–350 ms, 350–400 ms, and 400–420 ms showed shift in the distribution of electric potentials posterior regions to anterior regions of the brain in case of frequent stimuli. When the targets appeared, the potentials are seen to be widely distributed over posterior central regions and in case of distractors the potentials are observed to be localized over anterior frontal electrode sites which represented increase in cognitive demands while attention is laid during discrimination of the stimuli.

The reduced amplitude and increased latencies about 300 ms after the stimuli presentation are widely reported as abnormalities in schizophrenia as they index the amount of focus and time taken by the cognitive processes to classify the stimuli, respectively. Our results indicate increase in P300 amplitude and reduction in latency as vital sign of improvement in subjects enrolled in this experiment. The improvements can be attributed to prolonged treatments (both medication and psychosocial training routine) the patients have undergone since induction into the psychiatric setup.

One methodological constraint working with patients is to adapt the duration of the task according to the capacity of the patients to maintain participation in the experiment. The task was kept shorter which resulted into lower no of trials (160) for analysis in our case. The longer task durations are needed for better averaging and reduction of trial to trial variability. In future, work will be done with larger sample size and ERP components elicited from a longer duration task will be evaluated.

## References


1. Jutzeler, C.R., Mcmullen, M.E., Featherstone, R.F., Tatardleitman, V.M., Gandal, M.J., Carlson, G.C., Siegel, S.J.: Electrophysiological Deficits in Schizophrenia: Models and Mechanisms in Psychiatric Disorders Trends and Developments. IntechOpen (2011). ISBN 978-953-307-745-1
2. Oribe, N., Yoji, H., Kanba, S., Del Re, E., Seidman, L., Mesholam-Gately, R., Goldstein, J.M., et al.: Progressive reduction of visual P300 amplitude in patients with first-episode schizophrenia: an ERP study. Schizophr. Bull. **41**(2), 460–470 (2014)
3. Polich, J.: Updating P300: an integrative theory of P3a and P3b. Clin. Neurophysiol. **118**(10), 2128–2148 (2007)
4. Ford, J.M., Mathalon, D.H., Marsh, L., Faustman, W.O., Harris, D., Hoff, A.L., Beal, M., Pfefferbaum, A.: P300 amplitude is related to clinical state in severely and moderately ill patients with schizophrenia. Biol. Psychiatry **46**(1), 94–101 (1999)
5. Rao, K.J., Ananthnarayanan, C.V., Gangadhar, B.N., Janakiramaiah, N.: Smaller auditory P300 amplitude in schizophrenics in remission. Neuropsychobiology **32**(3), 171–174 (1995)
6. Turetsky, B.I., Cannon, T.D., Gur, R.E.: P300 subcomponent abnormalities in schizophrenia: III. Deficits in unaffected siblings of schizophrenic probands. Biol. Psychiatry **47**(5), 380–390 (2000)
7. Faux, S.F., McCarley, R.W., Nestor, P.G., Shenton, M.E., Pollak, S.D., Penhune,V., Mondrow, E., Marcy, B., Peterson, A., Horvath, T., Davis, K.L.: P300 topographic asymmetries are present in unmedicated schizophrenics. Electroencephalogr. Clin. Neurophysiol./Evoked Potentials Sect. **88**(1), 32–41 (1993)
8. Sahoo, S., Malhotra, S., Basu, D., Modi, M.: Auditory P300 event related potentials in acute and transient psychosis—comparison with schizophrenia. Asian J. Psychiatry **23**, 8–16 (2016)
9. McCathern, A.: Emitted P3a and P3b in chronic schizophrenia and in first episode schizophrenia. Doctoral dissertation, University of Pittsburgh. http://dscholarship.pitt.edu/31431/ (2017)
10. Kim, M., Lee, T.H., Kim, J.H., Hong, H., Lee, T.Y., Lee, Y., Salisbury, D.F., Kwon, J.S.: Decomposing P300 into correlates of genetic risk and current symptoms in schizophrenia: an inter-trial variability analysis. Schizophr. Res. (2017)
11. Vafaii, P., Mazhari, S., Pourrahimi, A.M., Nakhaee, N.: Hemispheric differences for visual P3 amplitude in patients with schizophrenia. Neuropsychiatry (London) **6**(6), 309 (2016)
12. Luck, S.J.: An Introduction to the Event-Related Potential Technique (Cognitive Neuroscience). The MIT Press (2005)
13. Jeek, P.: Database of EEG/ERP experiments: technical report no. DCSE/TR-201004. https://dspace5.zcu.cz/handle/11025/21566 (2010)
14. Acharya, J.N., Hani, A.J., Thirumala, P., Tsuchida, T.N.: American clinical neurophysiology society guideline 3: a proposal for standard montages to be used in clinical EEG. Neurodiagn. J. **56**(4), 253–260 (2016)
15. Delorme, A., Palmer, J., Onton, J., Oostenveld, R., Makeig, S.: Independent EEG sources are dipolar. PLoS ONE **7**(2), e30135 (2012)
16. Molina, G.N.G.: Direct brain-computer communication through scalp recorded EEG signals. École Polytechnique Fédérale de Lausanne (2004)
17. Chaumon, M., Bishop, D.V., Busch, N.A.: A practical guide to the selection of independent components of the electroencephalogram for artifact correction. J. Neurosci. Methods **250**, 47–63 (2015)
18. Mognon, A., Jovicich, J., Bruzzone, L., Buiatti, M.: ADJUST: An automatic EEG artifact detector based on the joint use of spatial and temporal features. Psychophysiology **48**(2), 229–240 (2011)

19. Delorme, A., Makeig, S.: EEGLAB: an open source toolbox for analysis of single-trial EEG dynamics including independent component analysis. J. Neurosci. Methods **134**(1), 9–21 (2004)
20. Lopez-Calderon, J., Luck, S.J.: ERPLAB: an opensource toolbox for the analysis of event-related potentials. Front. Hum. Neurosci. **8**, 213 (2014)
21. Picton, T.W.: The P300 wave of the human event related potential. J. Clin. Neurophysiol. **9**, 456–456 (1992)

# Analysis of Resting State EEG Signals of Adults with Attention-Deficit Hyperactivity Disorder

**Simranjit Kaur, Sukhwinder Singh, Priti Arun and Damanjeet Kaur**

**Abstract** Electroencephalography (EEG) has emerged as a valuable tool to understand the neurophysiology of Attention-Deficit Hyperactivity Disorder (ADHD) brain. The purpose of this work is to examine whether linear and nonlinear electrophysiological measures of adults with ADHD differ from control group during a rest state. To verify this, in the present study, EEG signals of 23 adults with ADHD and 27 control adults are recorded during 3 min eyes-open and eyes-closed conditions. Linear features are extracted from EEG epochs which include power spectra of delta, theta, alpha, beta, and gamma frequency bands. Nonlinear features are measured in terms of entropies to unveil signal complexity. Linear analysis showed that the ADHD group has increased power for slow waves and reduced power for fast waves. Nonlinear analysis results in significant reduction in approximate entropy, sample entropy, and Shannon entropy of the ADHD adults in comparison to control adults.

**Keywords** ADHD · Adults · EEG · Entropy · Rest state

## 1 Introduction

Attention-Deficit Hyperactivity Disorder (ADHD) is a pervasive neurodevelopmental condition defined in the presence of hyperactive, impulsive, and inattentive behavior that intervenes in normal functioning of the person. This disorder affects around 5% of the children worldwide [1]. ADHD can persist into adulthood in at least one-third of those with a diagnosis of ADHD in childhood [2]. An estimate of the worldwide prevalence rates of adult ADHD is found to be between 2.5 and 4.3%

S. Kaur · S. Singh (✉) · D. Kaur
University Institute of Engineering and Technology, Panjab University, Sector 25, Chandigarh, India
e-mail: sukhdalip@pu.ac.in

P. Arun
Government Medical College and Hospital, Chandigarh, India



R. Chaki et al. (eds.), *Advanced Computing and Systems for Security*, Advances in Intelligent Systems and Computing 897,
https://doi.org/10.1007/978-981-13-3250-0_5

[3–5]. Among this population, when ADHD students are analyzed, it is found that they report more depressive symptoms, intrusive thoughts, and task-interfering as compared to students without ADHD [6, 7]. These difficulties undermine their academic/occupational functioning in college settings that require organization, attention, and self-management and also increase their risk of substance abuse.

ADHD is a proven disorder in the field of medical practitioners. But with passing time, the use of neuroimaging techniques has provided a platform for better understanding the functioning of ADHD brain. Electroencephalography (EEG) attracted clinicians and researchers as it provides valuable information about high level of brain functioning. Electrophysiological traits have been used in research to study the heterogeneity of neural mechanisms of ADHD population. EEG measures have shown potential to serve as a neurometric tool and to increase reliability and validity of ADHD diagnosis. Previous EEG studies on ADHD children proved that there is increase in relative theta band power and decrease in relative alpha and beta band power during resting state condition [8, 9]. Theta/Beta ratio is also considered as a reliable EEG index in ADHD literature but it has more tendencies to act as prognostic measure rather than diagnostic measure [10]. Similar findings have been reported for ADHD adolescent population with the absolute dominance of delta and theta band power and a higher theta/beta ratio during an eyes-closed condition [11].

Few researchers examined EEG correlates of adults with ADHD in comparison to studies of children and adolescent ADHD and their results remain inconsistent. Bresnahan et al. [12] observed increased theta power in ADHD adults than healthy subjects but absolute beta and alpha band power is not significantly different in two groups. Koehler et al. [13] found increased absolute alpha and theta band power in ADHD adults but power density in delta and beta band is not different from control group. Clarke et al. [14] suggested that there is increase in the relative theta and decrease in the absolute delta and beta activity in ADHD adults. There are some procedural differences between these studies as EEG is recorded under eyes-open [12] and eyes-closed conditions [13, 14]. Woltering et al. [15] revealed that college students with ADHD have shown reduced power in fast frequencies and increased power for slow frequencies. These effects are more pronounced during eyes-closed in comparison to eyes-open baseline condition. Markovska-Simoska et al. [16] observed no significant differences between ADHD and control adults for spectral features extracted during an eyes-open state.

EEG signals express nonlinear interactions and dynamics especially in signals recorded during pathological disorders. Advancement of neural signal processing methods has enabled the use of various linear and nonlinear techniques to analyze complex and nonstationary EEG activity in epilepsy, anesthesia, and detection of sleep stages [17–20]. An extensive literature exists on spectral analysis of EEG signatures of ADHD subjects but the ability of nonlinear features has rarely been explored. Moreover, subpopulation of adult ADHD which enter in college life remains understudied in EEG research. In the current study, an attempt is made to explore several linear and nonlinear characteristics of brain signals of university students with ADHD and control adults. This analysis is done during two baseline conditions—eyes-closed and eyes-open. The objective of this research study is to examine differences between adults with ADHD and their non-ADHD peers in terms of their neural profile using spectral features and entropy-based measures. Linear analysis assumes stationarity of the underlying system but nonlinear measures capture statistical nonstationarity embedded in biomedical signals.

# 2 Methods and Materials

## *2.1 Participants*

A total of 50 adults, including 23 ADHD (*mean age* = 20.73) and 27 controls (*mean age* = 20.74), are recruited from a pool of university students. For identification of ADHD students, performance of Wender Utah Rating Scale (WURS) [21] to assess childhood ADHD symptoms and Adult ADHD Self Report Scale (ASRS) [22] to assess current symptoms of ADHD are filled by students. The students who have score above clinically significant levels are sent for further evaluation by psychiatrist to confirm the diagnosis. ADHD group must fulfill Diagnostic and Statistical Manual of Mental Disorders 5th edition (DSM-5) criteria of ADHD diagnosis. Mini International Neuropsychiatric Interview (MINI) version 6 is also filled by subjects to assess psychosocial comorbidities. Inclusion criteria are: (1) For ADHD group—a previous diagnosis of ADHD and existence of ADHD symptoms in adulthood as determined by clinical interview and ADHD rating scales (2) For control group—age-matched adults not having ADHD or any other psychopathology. Exclusion criteria are a comorbid current psychiatric diagnosis, substance dependence except Nicotine, head injuries, medical, or neurological disorder. This study is approved by the Institutional Research Ethics Board and all participants consented in written prior to the start of this study.

## *2.2 EEG Data Acquisition and Preprocessing*

The EEG data is recorded for ADHD and healthy volunteers under two different conditions: (i) 3-min eyes-open and (ii) 3-min eyes-closed. Participants are instructed to keep their eyes focused on a cross sign displayed on computer screen during an eyes-open state to prevent eye movements. Electrophysiological recordings are obtained from 19 scalp electrodes (Fp1, F7, F3, C3, P3, T3, T5, O1, Fp2, F8, F4, C4, P4, T4, T6, O2, Fz, Cz, and Pz) placed according to international 10/20 system [23]. Electrode impedance is maintained within similar range (5–10 K). All scalp electrodes are referenced to Cz electrode. The recording parameters are as sample rate of 256 Hz, bandpass hardware filter of 0.1–70 and 50 Hz notch filter. All data is exported to MATLAB (MathWorks, Natick, MA, USA) platform and re-referenced to average of all electrodes for further analysis. Electrooculographic (EOG) and Electromyogram

(EMG)-related artifactual components are removed from EEG data using an automated method proposed in [24]. Electrocardiography (ECG) related artifactual components are removed through Independent Component Analysis (ICA) decomposition in EEGLAB toolbox. Next, continuous EEG data are segmented into 2 s epochs separately for each of the two conditions and standard thresholding techniques are applied to reject artifacts. Epochs with extreme values, i.e., exceeding $\pm 75\ \mu V$ in any of the 19 channels at any time within the epoch are rejected. To detect linear drifts due to artifactual currents, data is fitted to a straight line and epoch is rejected if the slope exceeded a given threshold of 50 $\mu$V. Probability measure is employed to reject improbable epochs (represent artifact) with the help of standard deviation of mean probability distribution. Standard deviation limit set for single channel and for all channels is 5. A statistical measure kurtosis is used to reject data epochs with peaky activity value distribution and threshold of standard deviation is 5 for single channel as well as for all channels. Epochs with power spectra deviated from range [−100 25] dB in 20–40 Hz frequency window is also rejected as they might represent muscle activity. The average number of epochs left after artifact removal for ADHD group is 65.3 and 67.4 for control group under eyes-open condition. The average number of artifact-free epochs left for ADHD group is 73.1 and 75.4 for control group under eyes-closed condition.

### *2.3 Signal Feature Extraction*

**Linear Measures**. Spectral features consider signal in frequency domain and power spectrum of different frequency bands provide indication about person's brain states. Increased power in lower frequency oscillations during a rest state can reflect cortical slowing or drowsiness. Decreased power in high-frequency oscillations reflect reduced mental activity and concentration [25]. EEG signal is divided into five frequency bands, namely delta (0–4 Hz), theta (4–8 Hz), alpha (8–13 Hz), beta (13–30 Hz), and gamma (30–45 Hz) band using Discrete Wavelet Transform (DWT) [26]. The signal has been decomposed up to six levels with "sym8" as mother wavelet function. A Fast Fourier Transform (FFT) is performed separately on wavelet reconstructed signals obtained from approximation and detailed coefficients at different levels. It would find power spectral density for various frequency bands. FFT is defined as

$$X(k) = \sum x(n) w^{kn}, \text{ where } w = e^{-j2\pi/N}. \tag{1}$$

where x(n) is input signal, n is the number of input sample position, k is position of sample after FFT and N is maximum length of input signal. Absolute and relative powers have been calculated for all frequency bands.

**Nonlinear Measures**. Nonlinear methods of dynamics provide a useful set of tools to analyze EEG signals. They generate new information that linear measures cannot disclose, for example, about nonlinear interactions and the complexity and stability of underlying brain sites. Various entropy measures (like approximate entropy, sample entropy, Shannon entropy, and permutation entropy) have been used to characterize neural signals of ADHD and control adults. These features can reveal EEG desynchronization or interactions among multiple frequency components.

*Approximate Entropy (ApEn).* It can quantify system complexity and help in characterizing the chaotic behavior of time series [19]. ApEn (r, w, N) where two input arguments, a tolerance window w, and a run length r, must be given. N is the number of samples in time series. ApEn is computed as

$$\mathrm{ApEn}(r, w, N) = 1/(N - r) \sum \log\left(C^{r+1}(w)\right) - 1/(N - r + 1) \sum \log(C^{r}(w)). \tag{2}$$

where $C^r(w)$ is correlation integral. In the present study, the value of $r = 3$ and $w =$ 0.2 times the data standard deviation. The value of ApEn is high for complex or irregular time series as compared to regular or predictable data.

*Sample Entropy (SamEn).* It measures data regularity and self-similarity like ApEn [27]. However, SamEn is record length independent. It is the negative logarithm of the conditional probability that patterns of length m that match pointwise within a tolerance window w also match for length $m + 1$. Runs of points matching within the tolerance window w are carried out until there is no match. A(k) and B(k) are counters to store count of template matches for all lengths k up to m. Sample entropy is calculated as

$$\mathrm{SamEn}(k, w, N) = \ln(A(k)/B(k - 1)). \tag{3}$$

where $B(0) = N$, the length of time series and k is embedding dimension. The value of $k = 3$ and $w = 0.2$ times the data standard deviation. Larger values of SamEn correspond to more data irregularity and vice versa.

*Shannon Entropy.* It measures data uncertainty and is useful criteria to analyze and compare probability distribution [28]. Let S be a set of finite discrete random variables, $S = \{s_1, s_2, \ldots, s_m\}$, $s_m \in R^d$, Shannon entropy H(S) is defined as

$$H(S) = -k \sum p(s_i) \ln(p(s_i)). \tag{4}$$

where k is a positive constant and $p(s_i)$ is probability of $s_i \in S$ satisfying condition $\sum p(s_i) = 1$. High entropy indicates more chaotic system and less predictability.

*Permutation Entropy (PerEn).* This measure is based on permutation patterns analysis in arbitrary time series [29]. A set of $N - (m - 1)l$ vectors is generated for a time series $\{x_1, x_2, \ldots, x_N\}$ using embedding procedure. Here, l is lag, m is embedding dimension and a vector $X_t$ is defined by $[x_t, x_{t+\tau}, \ldots, x_{t+(m-1)l}]$. Each vector $X_t$ is then rearranged in an increasing order. There will be m! permutation patterns. Let $K(_i)$

denotes the number of occurrences of the permutation pattern $_i$ with i varies from 1 to m. Its relative frequency is obtained as $F(\pi) = K(\pi)/(N - (m - 1)1)$. The permutation entropy is described as

$$\mathrm{PerEn} = -\sum F(\pi)\ln(F(\pi)). \tag{5}$$

The value of PerEn ranges between 0 and log(m!) and its high value represents completely random time series. The selection of m value is important in the calculation of PerEn. The scheme will not work well for too small value of m (<3) as there would be only a few distinct states in signal. The condition $m! \leq N - (m - 1)l$ must hold to allow every possible permutation pattern to occur in a signal. The length of time series N must be larger than $m! + (m - 1)l$ to avoid undersampling. The value of m and l is chosen to be 5 and 1, respectively, in this study to satisfy this condition.

### 2.4 *Statistical Analysis*

The Quartile method is used to examine potential outliers in extracted features from data epochs. Separate one-way Analysis of Variance (ANOVA) is employed for each channel and measure to find out significant differences between ADHD and control group under eyes-open and eyes-closed condition. The level of significance $\alpha$ is set at 0.05.

## 3 Results and Discussion

### 3.1 *Linear Analysis*

Table 1 presents outcome of statistical test applied on spectral features extracted from different channels of ADHD and control group under eyes-open condition. Entries left blank in Table 1 indicate that no significant group difference is found corresponding to that particular feature and channel. One-way ANOVA results for eyes-open condition depicted significant effect of diagnosis (ADHD/Control) at Fp1 ($F(1,48) = 5.99$, $p = 0.0180$), and Fp2 ($F(1,48) = 17.17$, $p = 0.0001$) channel for absolute delta power. For absolute theta power, there is notable effect at Fp1 ($F(1,48) = 6.06$, $p = 0.0174$), and Fp2 ($F(1,48) = 12.55$, $p = 0.0008$) channel. There is significant effect of diagnosis for absolute alpha power at Fp1 ($F(1,48) = 5.72$, $p = 0.0207$) and Fp2 ($F(1,48) = 13.49$, $p = 0.0006$). Absolute and relative beta power shown significant differences between two groups at Fp2 ($F(1,48) = 4.47$, $p = 0.0397$) and C4 ($F(1,48) = 5.98$, $p = 0.0181$) channel, respectively. Relative gamma power displayed pronounced effect at Fz ($F(1,48) = 4.51$, $p = 0.0389$), C4 ($F(1,48) = 5.81$, $p = 0.0198$) and Pz ($F(1,48) = 4.5$, $p = 0.0389$) channel. Absolute gamma, relative delta, relative

**Table 1** Statistical analysis of spectral features under eyes-open condition

| Channel | Abs delta power | Abs theta power | Abs alpha power | Abs beta power | Rel beta power | Rel gamma power |
|---|---|---|---|---|---|---|
| 1. Fp1 | F=5.99, p=2e−02 | F=6.06, p=2e−02 | F=5.72, p=2e−02 | – | – | – |
| 2. Fp2 | F=17.17, p=1e−04 | F=12.55, p=8e−04 | F=13.49, p=6e−04 | F=4.47, p=4e−02 | – | – |
| 3. Fz | – | – | – | – | – | F=4.51, p=4e−02 |
| 4. C4 | – | – | – | – | F=5.98, p=2e−02 | F=5.81, p=2e−02 |
| 5. Pz | – | – | – | – | – | F=4.5, p=4e−02 |

theta, and relative alpha do not differ significantly between two groups under eyes-open state.

Table 2 presents the outcome of statistical test applied on spectral features extracted from different channels of ADHD and control group under eyes-closed condition. This analysis resulted in notable effect of diagnosis at Fp2 (F(1,48)=7.5, p=0.0087), F8 (F(1,48)=8.21, p=0.0061), F7 (F(1,48)=8.29, p=0.0059), C4 (F(1,48)=6.5, p=0.0140), Pz (F(1,48)=6.94, p=0.0113), T5 (F(1,48)=8.77, p=0.0047), and O1 (F(1,48)=7.79, p=0.0075) channel for relative delta power and at C4 (F(1,48)=4.14, p=0.0473) for absolute delta power. There are significant group differences for relative theta at F4 (F(1,48)=6.96, p=0.0112), C4 (F(1,48)=4.21, p=0.0457), and T4 (F(1,48)=5.46, p=0.0237). There is a significant effect of diagnosis for relative beta power at Fp2 (F(1,48)=4.21, p=0.0456), C4 (F(1,48)=7.98, p=0.0068), P3 (F(1,48)=4.66, p=0.0358), Pz (F(1,48)=4.78, p=0.0337), T5 (F(1,48)=7.51, p=0.0086), and O2 (F(1,48)=7.13, p=0.0103) channel and for absolute beta power at O2 (F(1,48)=5.95, p=0.0185). There is pronounced effect of diagnosis for relative gamma power at Fp2 (F(1,48)=7.05, p=0.0105), F7 (F(1,48)=4.12, p=0.0481), C4 (F(1,48)=6.62, p=0.0132), Pz (F(1,48)=6.54, p=0.0137), and T3 (F(1,48)=4.59, p=0.0373). Absolute theta, absolute alpha, absolute gamma, and relative alpha power do not differ significantly between two groups under eyes-closed state. Figure 1 and Fig. 2 presents comparison of absolute power spectra of ADHD and control group under eyes-open and eyes-closed conditions at Fp2 and C4 channel, respectively.

## 3.2 Nonlinear Analysis

Analysis of entropy-based measures under eyes-open condition depicted significant differences between ADHD and control group at Fp2 ($F(1,48) = 12.89$, $p = 0.0007$) channel for approximate entropy. There are also notable differences between two

**Table 2** Statistical analysis of spectral features under eyes-closed condition

| Channel | Abs delta power | Abs beta power | Rel delta power | Rel theta power | Rel beta power | Rel gamma power |
|---|---|---|---|---|---|---|
| 1. Fp2 | – | – | F=7.5, p=8e−03 | – | F=4.21, p=4e−02 | F=7.05, p=1e−02 |
| 2. F4 | – | – | – | F=6.96, p=1e−02 | – | – |
| 3. F7 | – | – | F=8.29, p=6e−03 | – | – | F=4.12, p=5e−02 |
| 4. F8 | – | – | F=8.21, p=6e−03 | – | – | – |
| 5. C4 | F=4.14, p=5e−02 | – | F=6.5, p=1e−02 | F=4.21, p=4e−02 | F=7.98, p=6e−03 | $F$ =6.62, $p$ =1e−02 |
| 6. P3 | – | – | – | – | F=4.66, p=3e−02 | – |
| 7. Pz | – | – | F=6.94, p=1e−02 | – | F=4.78, p=3e−02 | $F$ =6.54, $p$ =1e−02 |
| 8. T3 | – | – | – | – | – | $F$ =4.59, $p$ =4e−02 |
| 9. T4 | – | – | – | F=5.46, p=2e−02 | – | – |
| 10. T5 | – | – | F=8.77, p=5e−03 | – | F=7.51, p=8e−03 | – |
| 11. O1 | – | – | F=7.79, p=7e−03 | – | – | – |
| 12. O2 | – | F=5.95, p=2e−02 | – | – | F=7.13, p=1e−02 | – |

groups for sample entropy at Fp2 ($F$(1,48) = 12.86, $p$ = 0.0007) channel. Shannon entropy showed pronounced effect of diagnosis (ADHD/Control) at Fp1 ($F$(1,48) = 10.15, $p$ = 0.0025) and Fp2 ($F$(1,48) = 14.43, $p$ = 0.0004) channel. There are no significant group differences at any channel for permutation entropy in eyes-open condition. Analysis under eyes-closed condition resulted notable effect of diagnosis at O2 ($F$(1,48) = 4.19, $p$ = 0.0461) channel for Shannon entropy. There are no significant group differences at any channel for permutation entropy in eyes-closed condition. Figure 3 presents comparison of approximate and sample entropy of ADHD and control group under eyes-open condition at Fp2 channel.

## 4 Conclusion

In the present study, spectral and entropy-based measures obtained from brain signals of ADHD and control adults under rest state conditions are examined in order to discriminate clearly between two groups. The ADHD group exhibited distinct EEG

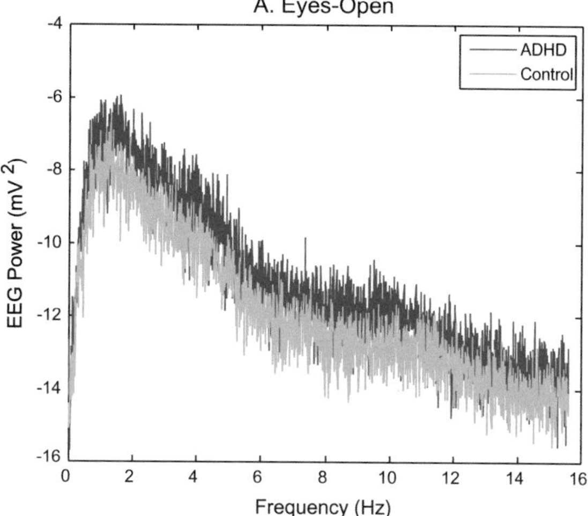


**Fig. 1** Power spectra of ADHD and control group under eyes-open condition at Fp2 channel

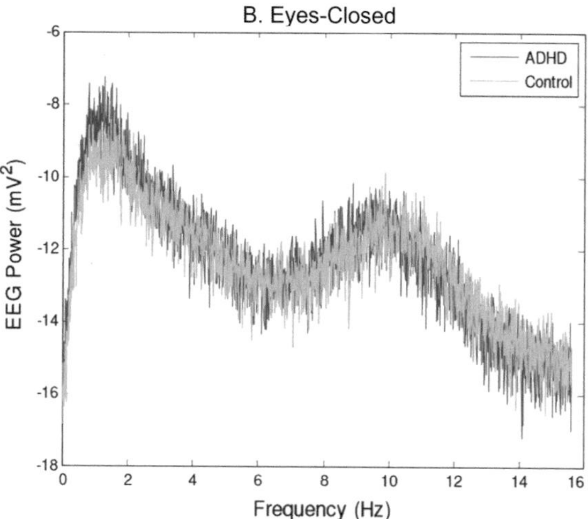


**Fig. 2** Power spectra of ADHD and control group under eyes-closed condition at C4 channel

profile under eyes-open and eyes-closed conditions. Under eyes-open condition, it has been found that absolute delta and theta band power is high for ADHD group as compared to normal adults at Fp1 and Fp2 electrode locations. Absolute alpha power is higher for ADHD adults than normal subjects at FP1 and Fp2 channels. The ADHD group has lower relative beta power than control adults at C4 channel but absolute beta power is found to be high at FP2 channel for ADHD adults. Relative gamma power is decreased in ADHD group at Fz, C4 and Pz channels under eyes-open condition. Difference between ADHD and control group is large in terms of mean absolute delta, theta and alpha band power at Fp2 channel. Under eyes-closed condition, it has been found that ADHD adults have high absolute delta power at C4 channel and

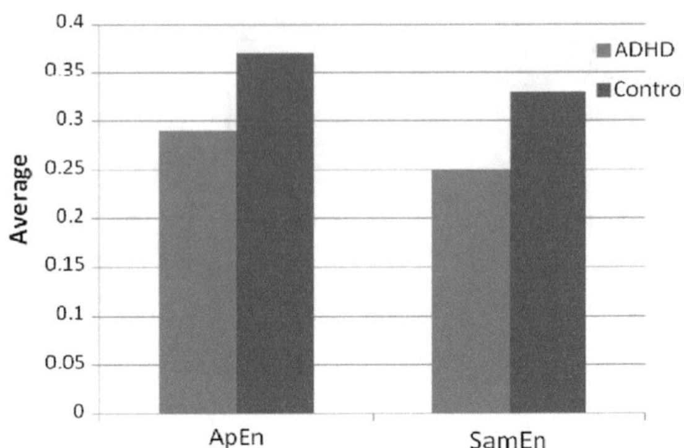


**Fig. 3** Comparison of approximate and sample entropy of ADHD and control group

high relative delta power at F7, F8, C4, Pz, and O1 channels. Relative delta power of ADHD group is low at Fp2 and T5 channels. Relative theta band power is low for ADHD at F4, C4, and T4 channels. The ADHD group has lower relative beta power than control adults at Fp2, C4, Pz, P3, T5, and O2 channels and lower absolute beta power at O2 channel. Relative gamma power is decreased at Fp2, F7, C4, Pz, and T3 channels in ADHD group in eyes-closed condition. Significant group differences are obtained at multiple channels under eyes-closed as compared to eyes-open condition for spectral features. Relative power spectra is helpful under eyes-closed and absolute power based features are of use under eyes-open to make a clear distinction between two groups. There is increased alpha power for both groups under eyes-closed in comparison to eyes-open condition. Increased slow waves power and reduced fast band powers in ADHD adults can be explained by hypoaroused model of ADHD [8]. This model proposes that cortical activity of ADHD people is below an optimal level. There is an association between EEG abnormalities (increased theta and decreased beta power) and hypoarousal.

Under eyes-open condition, it has been found that approximate entropy is low for ADHD group as compared to normal adults at Fp2 electrode location. Similarly, sample entropy has lower values for ADHD adults than normal adults at FP2 channel. The ADHD group has reduced Shannon or spectral entropy than control adults at Fp1 and Fp2 channels. It has been seen that there is large difference between ADHD and control adults in respect of entropy-based measures at Fp2 channel. Under eyes-closed condition, it has been found that the ADHD group has increased Shannon or spectral entropy than control adults at O2 channel. These results indicate that signals of ADHD adults exhibited less chaotic behavior as compared to normal adults under eyes-open condition. It has been observed that entropy based measures and absolute power based features are more reliable under eyes-open for group differentiation as their $p$-values of statistical ANOVA test are much lower than 0.05. Relative power-based features can be employed as discriminators under eyes-closed state. This study suggests that nonlinear EEG measures can be a useful index under eyes-open condition because their findings remain consistent across channels, unlike linear measures.

## 5 Future Scope

In addition to resting state, analysis of EEG data under different cognitive conditions might be useful in understanding the neural mechanisms of adults with ADHD. Other time domain and nonlinear signal measures need to be explored in future to assess their diagnostic ability. The use of advanced statistical methods like graph theory, machine learning, logistic regression, and discriminant functions would be helpful to model complex relationships in the EEG data.

## References


1. Polanczyk, G., de Lima, M.S., Horta, B.L., Biederman, J., Rohde, L.A.: The worldwide prevalence of ADHD: a systematic review and metaregression analysis. Am. J. Psychiatry **164**(6), 942–948 (2007)
2. Wender, P.H., Wolf, L.E., Wasserstein, J.: Adults with ADHD. Ann. N. Y. Acad. Sci. **931**(1), 1–16 (2001)
3. Simon, V., Czobor, P., Bálint, S., Mészáros, Á., Bitter, I.: Prevalence and correlates of adult attention-deficit hyperactivity disorder: meta-analysis. Br. J. Psychiatry **194**(3), 204–211 (2009)
4. Fayyad, J., De Graaf, R., Kessler, R., Alonso, J., Angermeyer, M., Demyttenaere, K., Lépine, J.P.: Cross-national prevalence and correlates of adult attention-deficit hyperactivity disorder. Br. J. Psychiatry **190**(5), 402–409 (2007)
5. Kessler, R.C., Adler, L., Barkley, R., Biederman, J., Conners, C.K., Demler, O., Spencer, T.: The prevalence and correlates of adult ADHD in the United States: results from the National Comorbidity Survey Replication. Am. J. Psychiatry **163**(4), 716–723 (2006)
6. Rabiner, D.L., Anastopoulos, A.D., Costello, J., Hoyle, R.H., Swartzwelder, H.S.: Adjustment to college in students with ADHD. J. Atten. Disord. **11**(6), 689–699 (2008)
7. Shaw-Zirt, B., Popali-Lehane, L., Chaplin, W., Bergman, A.: Adjustment, social skills, and self-esteem in college students with symptoms of ADHD. J. Atten. Disord. **8**(3), 109–120 (2005)
8. Barry, R.J., Clarke, A.R., Johnstone, S.J.: A review of electrophysiology in attention-deficit/hyperactivity disorder: I. Qualitative and quantitative electroencephalography. Clin. Neurophysiol. **114**(2), 171–183 (2003)
9. Snyder, S.M., Hall, J.R.: A meta-analysis of quantitative EEG power associated with attention-deficit hyperactivity disorder. J. Clin. Neurophysiol. **23**(5), 441–456 (2006)
10. Arns, M., Conners, C.K., Kraemer, H.C.: A decade of EEG theta/beta ratio research in ADHD: a meta-analysis. J. Atten. Disord. **17**(5), 374–383 (2013)
11. Hobbs, M.J., Clarke, A.R., Barry, R.J., McCarthy, R., Selikowitz, M.: EEG abnormalities in adolescent males with AD/HD. Clin. Neurophysiol. **118**(2), 363–371 (2007)
12. Bresnahan, S.M., Barry, R.J., Clarke, A.R., Johnstone, S.J.: Quantitative EEG analysis in dexamphetamine-responsive adults with attention-deficit/hyperactivity disorder. Psychiatry Res. **141**(2), 151–159 (2006)
13. Koehler, S., Lauer, P., Schreppel, T., Jacob, C., Heine, M., Boreatti-Hümmer, A., Herrmann, M.J.: Increased EEG power density in alpha and theta bands in adult ADHD patients. J. Neural Transm. **116**(1), 97–104 (2009)
14. Clarke, A.R., Barry, R.J., Heaven, P.C., McCarthy, R., Selikowitz, M., Byrne, M.K.: EEG in adults with attention-deficit/hyperactivity disorder. Int. J. Psychophysiol. **70**(3), 176–183 (2008)
15. Woltering, S., Jung, J., Liu, Z., Tannock, R.: Resting state EEG oscillatory power differences in ADHD college students and their peers. Behav. Brain Funct. **8**(1), 60–68 (2012)

16. Markovska-Simoska, S., Pop-Jordanova, N.: Quantitative EEG in children and adults with attention deficit hyperactivity disorder: comparison of absolute and relative power spectra and theta/beta ratio. Clin. EEG Neurosci. **48**(1), 20–32 (2017)
17. Acharya, U.R., Sree, S.V., Swapna, G., Martis, R.J., Suri, J.S.: Automated EEG analysis of epilepsy: a review. Knowl.-Based Syst. **45**, 147–165 (2013)
18. Al-Kadi, M.I., Reaz, M.B.I., Ali, M.A.M.: Evolution of electroencephalogram signal analysis techniques during anesthesia. Sensors **13**(5), 6605–6635 (2013)
19. Acharya, R., Faust, O., Kannathal, N., Chua, T., Laxminarayan, S.: Non-linear analysis of EEG signals at various sleep stages. Comput. Methods Programs Biomed. **80**(1), 37–45 (2005)
20. Artameeyanant, P., Chiracharit, W., Chamnongthai, K.: Spike and epileptic seizure detection using wavelet packet transform based on approximate entropy and energy with artificial neural network. In: Biomedical Engineering International Conference (BMEiCON), 2012, pp. 1–5. IEEE (2012)
21. Ward, M.F.: The Wender Utah Rating Scale: An aid in the retrospective diagnosis of childhood attention deficit hyperactivity disorder. Am. J. Psychiatry **150**, 885–990 (1993)
22. Kessler, R.C., Adler, L., Ames, M., Demler, O., Faraone, S., Hiripi, E.V.A., Ustun, T.B.: The World Health Organization Adult ADHD Self-Report Scale (ASRS): a short screening scale for use in the general population. Psychol. Med. **35**(2), 245–256 (2005)
23. Jasper, H.: Report of the committee on methods of clinical examination in electroencephalography. Electroencephalogr. Clin. Neurophysiol. **10**, 370–375 (1958)
24. Daly, I., Scherer, R., Billinger, M., Müller-Putz, G.: FORCe: Fully Online and automated artifact Removal for brain-Computer interfacing. IEEE Trans. Neural Syst. Rehabil. Eng. **23**(5), 725–736 (2015)
25. Loo, S.K., Makeig, S.: Clinical utility of EEG in attention-deficit/hyperactivity disorder: a research update. Neurotherapeutics **9**(3), 569–587 (2012)
26. Wali, M.K., Murugappan, M., Ahmmad, B.: Wavelet packet transform based driver distraction level classification using EEG. Math. Problems in Eng. **2013** (2013)
27. Richman, J.S., Moorman, J.R.: Physiological time-series analysis using approximate entropy and sample entropy. Am. J. Physiol. Heart Circ. Physiol. **278**(6), H2039–H2049 (2000)
28. Shannon, C.E.: A mathematical theory of communication. ACM SIGMOBILE Mob. Comput. Commun. Rev. **5**(1), 3–55 (2001)
29. Bandt, C., Pompe, B.: Permutation entropy: a natural complexity measure for time series. Phys. Rev. Lett. **88**(17), 174102–174106 (2002)

# Readability Analysis of Textual Content Using Eye Tracking

**Aniruddha Sinha, Rikayan Chaki, Bikram De Kumar and Sanjoy Kumar Saha**

**Abstract** Characterization of silent reading involves a joint analysis of the reading material and reader's capacity to assimilate the content. The amount of mental workload, to understand a content, reflects the overall cognitive load which is commonly measured using electrophysiological signals. Eye movement provides a direct means of evaluating one's reading characteristics in much finer details. In this paper, we use a commercially available low-cost eye-tracking device, EyeTribe. We analyze the eye movement behavior to compare readability of textual contents and also characterize an individual's reading profile. The experiment is performed using two types of textual contents significantly separated in difficulty level as evaluated using standard readability indices. The Flesch-Kincaid Grade Level for one type varies between 3 and 6 and the other between 13 and 16. Eye gaze analysis is done on features related to fixation and saccade using both global and local features. Results indicate the difference in content is reflected in global features and individual level variations, for a given type of content, are observed in the entropy derived from local features.

**Keywords** Readability analysis · Eye gaze · Fixation · Entropy · Learning

A. Sinha (✉)
TCS Research & Innovation, Tata Consultancy Services, Kolkata, India
e-mail: aniruddha.s@tcs.com

R. Chaki · B. De Kumar · S. K. Saha
Department of Computer Science & Engineering, Jadavpur University, Kolkata, India
e-mail: rikayan@acm.org

B. De Kumar
e-mail: bikramkumarde@yahoo.in

S. K. Saha
e-mail: sks_ju@yahoo.co.in



R. Chaki et al. (eds.), *Advanced Computing and Systems for Security*,
Advances in Intelligent Systems and Computing 897,
https://doi.org/10.1007/978-981-13-3250-0_6

## 1 Introduction

With the advancement in technology, the education system is undergoing a transformation in terms of the creation of training material and interaction between teacher and student. Recently, distant learning [1] has gained tremendous popularity which has enabled need-based augmentation of knowledge for both students and professionals. In order to enable personalization in the e-learning scenario, there is a need for matchmaking between the learner's profile and the educational content. This demands for an automatic evaluation of educational content coupled with profiling of reading characteristics of individuals.

There exist few standard methods to evaluate a training material. The textual content is evaluated based on Natural Language processing (NLP) by deriving readability indices, namely Flesch-Kincaid Reading Ease [2], Flesch-Kincaid Grade Level, SMOG Index, Coleman Liau Index, etc. There are statistical methods [3] to evaluate a text in terms of number average words per sentence, average syllables per word, density of complex words, etc. There are evaluations in terms of graphical representation of information where linear is preferred against radial graphs [4]. Feedback based methods give a gross level evaluation of the content and has individual biases [5]. The principles behind the effect of learning using multimedia content are studied by Meyer [6] where a major emphasis is given in the cognitive load of a content.

The cognitive load [7] is the amount of mental workload experienced, by the working memory, while performing a task. Various physiological sensing like Electroencephalogram (EEG) [8], Galvanic Skin Response (GSR) [9], Electrocardiogram (ECG) [10], skin temperature [11], Electromyogram (EMG) [12], Photoplethysmogram (PPG) [13], etc., have been used to determine the cognitive workload. These methods require wearing of certain sensors and provide a gross information on the cognitive load. On the other hand behavior of eye gaze and pupil dilation during a reading task provides a direct mechanism to quantify the attention, engagement, flow of the reading and the cognitive load in more granular manner [14]. Analysis of silent reading using eye tracking can serve two purposes—evaluation of a content for a given category of readers, analysis of the reader for a given content. Depending on the level of the reader, the content which is most suitable can be selected for consumption.

Content analysis using high-end Infrared (IR) based eye trackers provides promising results. Reney et al. [15] used EyeLink 1000 along with a chin rest, to analyze fixations to evaluate the cognitive processed in the text comprehension. Navarro et al. [16] analyzed the effect of highlighting contents using colors with a cost eye-tracking device, Tobii X60. This work was motivated design principles of signaling mechanism, supporting cues, contrast to distinguish the effect of tone in colors [17]. Eye-tracking systems also find applications in domains of medicine, Human–Computer Interactions (HCI), biological engineering and psychology. Burton et al. [18] and Maruta et al. [19] have used silent reading task to evaluate patients with Glaucoma. Linda et al. [20] analyzed the effect of eye movement due to malnutrition in children using Tobii X2-60 eye tracker. However, all these works are targeted toward evaluation of content using very costly eye-tracking devices which is not realistic in the e-learning scenario. Recent publications [21] on practical guide to use eye trackers during reading also analyze the performance using devices with sampling rate of 500 and 1000 Hz.

Given the limited amount of study on characterization of silent reading using low-cost eye tracker devices, coupled with challenges due to the high noise characteristics [22], we decided to evaluate the feasibility of the same. In this paper, we limit ourselves on the eye gaze movement behavior and derive features related fixations and saccades [23] using a low-cost eye-tracking device, EyeTribe [24]. The overall difficulty level of a reading content is measured using global features and the individual reading characteristics are derived using the entropies of the local line level features.

The novelty of this paper lies in finding the relation between the fixation duration and transition frequency between the fixation and saccades for easy and difficult content. These features provide gross level separation between the two types of content and also give valuable insights into participants' reading behavior. Additionally, the entropies derived from the above features provide a good measure for a content in terms of the easiness of consumption.

The paper is organized as follows. Section 2 gives the detailed description of experimental setup and stimuli used in this paper. Section 3 describes the methodology of processing the eye tracker signal and extraction of features. Section 4 presents the experimental results followed by conclusion in last section.

## 2 Experimental Paradigm

In this section we explain the stimulus used, experimental setup, and the data capture procedures in detail.

### *2.1 Description of the Stimulus*

The stimuli are two sets of paragraphs, one relatively easy compared to the other. Each set containing three paragraphs, each having 12–14 lines and 132–164 words. The easy paragraphs are constructed with information related to the atmosphere of moon and sun; life of a hawker; life of a farmer. The difficult paragraphs are constructed based on critics on the novel "Great Expectations" of Charles Dickens; recent effects of demonetization; current Indian political scenario. To justify the difference between the two types of the texts, NLP based readability analysis and statistical evaluation are performed as shown in Table 1 and Table 2 respectively. For easy contents, the metric "Flesch Kincaid Reading Ease" is much higher in the range of 73–93 whereas for difficult contents it is in the range of 28–40. The other NLP parameters are much

**Table 1** NLP based readability analysis of the contents (three each for easy and difficult)

| Parameter | Easy | Difficult | Interpretation |
|---|---|---|---|
| Flesch-Kincaid Reading Ease (0–100) | 93.8, 85, 73.2 | 28.7, 33.4, 40.8 | Above 40 anyone can understand, lower values indicate that the text is more difficult to comprehend |
| Flesch-Kincaid Grade Level (>3) | 3.1, 4.4, 6.2 | 15.9, 14.9, 13.3 | Educational grade level required to understand the text. The lower the grade, the more readable the text |
| SMOG Index (4–18) | 4.2, 5.8, 6.2 | 12.4, 12.9, 11.8 | Higher values indicate that the text is more difficult |
| Coleman Liau Index (>1) | 6.2, 9.1, 10.3 | 15.4, 14.7, 12.2 | Educational grade level required to understand the text. The lower the grade, the more readable the text |

**Table 2** Statistical analysis of the contents (three each for easy and difficult)

| Parameter | Easy | Difficult |
|---|---|---|
| Number of sentences | 14, 12, 13 | 5, 6, 7 |
| Number of words | 162, 146, 163 | 134, 151, 162 |
| Percentage of complex words | 3.70, 7.53, 8.59 | 17.16, 19.87, 17.9 |
| Average words per sentence | 11.57, 12.17, 12.54 | 26.80, 25.17, 23.14 |
| Average syllables per word | 1.2, 1.29, 1.43 | 1.78, 1.75, 1.69 |

higher for difficult contents as compared to easy contents. Table 1 gives the ranges and implications of these parameters [25].

From Table 2, it can be seen the range of words are very similar for both types of contents however, the higher density of complex words, long sentences and higher average syllables per word differentiates the difficult contents from the easy ones. Moreover, feedback from the participants in the Likert scale [26] of 5, eight out of nine participants indicated 5 (strongly agree) and one indicated 4 (agree).

## 2.2 *Participants*

Thirteen participants (eight males and five females, mean age: 33 $\pm$ 12 years) from an Engineering Institute participated in the study. We ensured that they have similar cultural and educational backgrounds. All the participants have normal or corrected to normal vision with spectacles and all are right handed. At the beginning, the task and the procedure were explained to them. In order to ensure that the participants followed the tasks properly, a demo version of the task is performed before starting the actual data capture. After the experiment, feedback is taken from the participants in a 5-point Likert scale [3], on whether they regularly read English novels, find the text difficult or easy, had to fix on many words for long duration, had to retrace texts during reading, the enjoyment and concentration level during reading.

## 2.3 Experimental Setup

We have used EyeTribe eye tracker for the study. The sampling frequency of the device is 30 Hz. An initial calibration is done using the SDK based calibration procedure to obtain a calibration score of 5, which is indicative of good calibration. As we are interested to study the eye motion during reading task, the error in the target gaze location is important. Hence, the systematic error [27] is reduced using a good calibration for all the participants. The experimental setup is shown in Fig. 2.

The eye tracker is placed on a small tripod is in front of a 21-in. Liquid Crystal Display (LCD) monitor. The LCD screen (1600 $\times$ 1200 pixels resolution) is placed approximately 60 cm away from the participant (Fig. 1). The experiment is performed without any chin rest to emulate the real scenario. However participants are asked to minimize their head movement. The experiment is carried out in a closed quite room with constant lighting conditions. The brightness and the contrast of the LCD are adjusted based on the comfort level of the participants. The text is displayed in 32-point Calibri font, in black letters on a white background, in which each letter subtends a maximum height of 0.56° visual angle. Participants are asked to perform silent reading while understanding the meaning of the content, as quickly and accurately as possible, similar to reading a novel or paper.

## 2.4 Reading Experiment

During the experiment, the texts for each paragraph are kept fixed and non-scrolling. Initially a 10 s baseline section preceded the task during which the participants are asked to focus on the fixation cross in white background that is displayed in the midst of the computer screen. After that textual contents are shown one after another. Once the participant finishes reading the first content, the next button is pressed to go the next content as shown in Fig. 2. In this manner, three same types (easy or difficult) of contents are read. Then a 5 min break is given the subject to get back to relaxed condition. After that, again a 10 s baseline followed by three texts of different types is read. The sequence of easy and difficult texts is randomized to get rid of any bias.

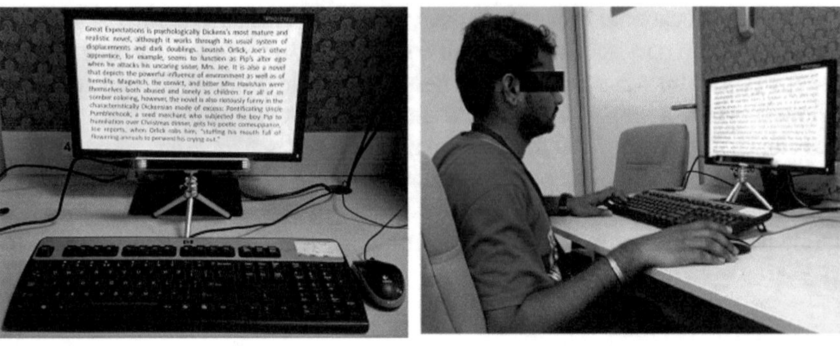

**Fig. 1** Experimental Setup—**a** with display, eye tracker, keyboard and mouse, **b** view with participant using the setup

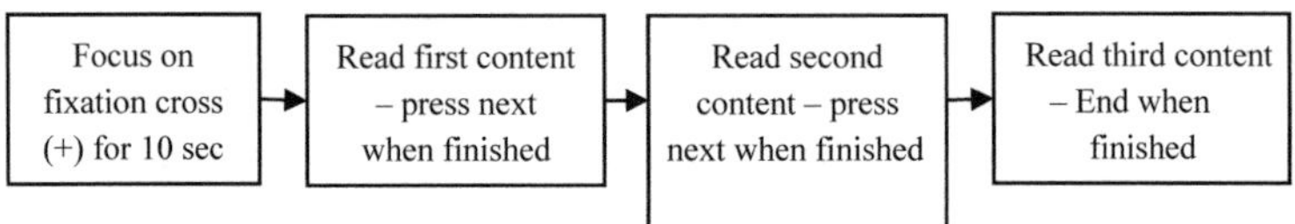


**Fig. 2** Sequence of the stimuli as shown on the display screen

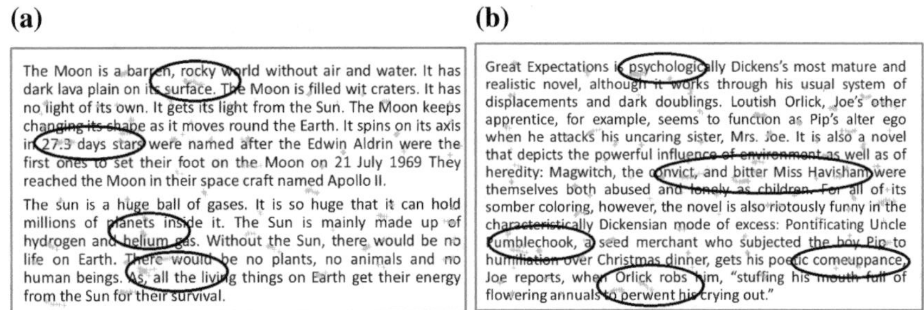


**Fig. 3** Text overlaid with eye tracker fixation information (**a**) an easy, **b** a difficult content

Two sample contents along with the fixations (concentrated green) as derived from eye tracker are shown in Fig. 3. It can be seen that for easy text (Fig. 3a) there is few fixations especially for the words "*rocky*", "*helium*" and a number ("*27.3*"). Apart from that there is another fixation at the beginning of the last sentence where it starts with "*As, all …*", may be due to continuing context from the previous line. For difficult content (Fig. 3b), the fixations are majorly for the difficult words (e.g., "*psychologically*", "*comeuppance*") and proper nouns with higher duration indicated by the amount of green points.

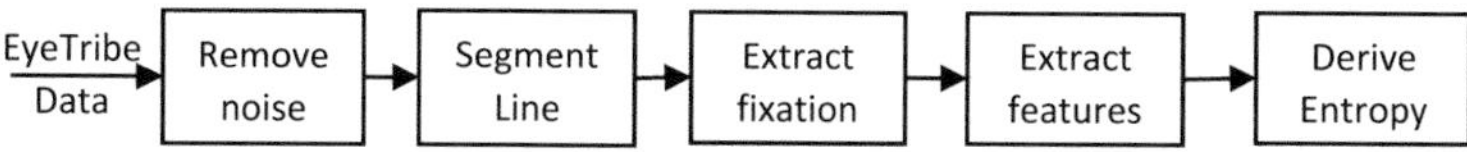


**Fig. 4** Block diagram for the proposed methodology

# 3 Methodology

During the silent reading task for the stimuli described in Sect. 2.1, the eye gaze data is captured using EyeTribe. This is a time series of X and Y coordinates of the display screen where a participant is looking at and is captured at 30 Hz for each eye. This data is used to derive the saccades and fixations.

Saccades are commonly observed during a search tasks. A jerky movement occurs in both eyes for about 2–10° of visual angle, and duration is about 25–100 ms. The approximate rotational velocity is 500–900 °/s having very high acceleration [28].

Each saccade is followed by a fixation, where the eye has a 250–500 ms dwell to process visual information [23]. These saccade-fixation sequences form scan paths, providing a rich set of data for tracking visual attention on a display [29]. Fixation followed by a saccade while reading forward is called forward fixation and fixation followed by a saccade that moves the reader to behind in a text is called regressive fixation [30].

Smooth Pursuit eye movements are much slower (1–30 °/s) movements used to track slowly and regularly moving visual targets. These cannot be voluntarily generated, and may be independently superimposed upon saccades to stabilize moving targets on the retina [31].

In the case of silent reading, search is done for the next word and hence the saccadic eye movements and fixations are the primary components. Due to the low-cost hardware the noise associated with the time series X–Y coordinates of the eye gaze data is cleaned using a preprocessing technique. Time segmentation of the signal is performed to extract the data corresponding to each line and then derive the local and global features. Global features are used to compare the different type of contents and the entropies derived from the local features provide individual characteristics of the participants. A flowchart for the overall methodology is shown in Fig. 4.

## *3.1 Noise Removal*

The raw eye gaze data from EyeTribe contains information about the state of the signal quality in the metadata. In each sample of eye gaze data, state indicates whether the sample contains X–Y coordinates for both the eye or not. Sometimes data is only available for one eye or sometimes for both of them. This is due to the spurious noise occurring from the IR interference in the environment. Other factors for the noise include eye blinks and momentary head movements. In all the above scenarios the

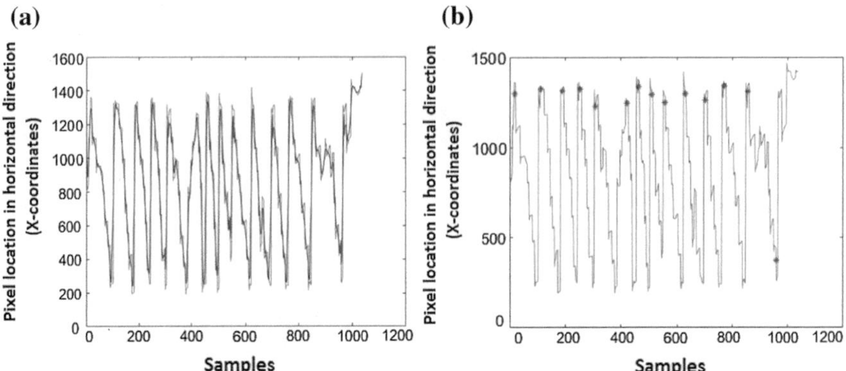


**Fig. 5** Filtering and Segmentation of line using peak detection. For participant-1 while reading first easy text, **a** the raw eye gaze data (blue) and low pass filtered output (red), **b** the raw eye gaze data (blue) and the detected peaks (red "*")

eye gaze data is interpolated using standard bi-cubic interpolation of the adjacent good data. This interpolated eye gaze data is then used for further processing.

## *3.2 Segmentation of Line*

When the text is read, for each line, the eye gaze moves from left to right and then jumps to the left of screen to go to the next line. Each text on the screen consists of 13–15 lines. Hence there is a large saccadic movement from right to left at the end of each line. This is reflected as an oscillating signal in the X-coordinate of the eye-tracking data. The segment between two peaks correspond to the segments of the line in time domain. Even though the preprocessing of raw signal is done for feature extraction, it is quite noisy for correctly detecting the peaks that would correspond to the segments of line. In order to detect the segments of the eye-tracking signal that corresponds to each line of the text, initially the finer changes in the eye gaze data is removed using filter method. A low pass filtering is performed with a finite impulse response (FIR) filter of 15 taps. The filter tap weights are derived using Hanning window [32]. Peak detection [33] is performed on the filtered data to derive the sample points (i.e., the timestamps) of the start and end of the lines. In Fig. 5, the variation of the X-coordinate are shown in blue line and the detected peaks are shown in red star "*", indicating the end of the line.

The sample plots of eye gaze data for participant 1 while reading the first easy content are shown after for low pass filtering and peak (line) detection in Fig. 5a and b, respectively. Sample plots for participant 2 for first easy content and third difficult content are shown in Fig. 5c and d, respectively. It can be seen that for all the scenarios the peaks are detected quite well. In the case of Fig. 5c and d there are certain premature peaks indicating retrace of the text after reading a partial line. The eye-tracking data derived from these line segments are then used to derive local features for each line.

### 3.3 Feature Extraction

In order to analyze a content and its reading behavior, some of the frequently used standard methods are related to blink rates [4, 14] and duration of the reading. Apart from these, the behavior related to fixation and saccades can provide information on how smoothly a content can be read. Thus initially, local features are extracted from each line and then an overall information is derived for the content by taking the aggregation of the line features, termed as global features. The global features namely, *Normalized Fixation Duration* and *Normalized Fixation Switching* are used to analyze the type of content (easy or difficult). During reading if there is a difficult word or phrase then it is expected to have longer duration fixation in the eye gaze. Thus more the value of *Normalized Fixation Duration*, the more is the experienced cognitive load. On the other hand for an easy text, it is expected that the reader would read the content quite fast and there would be large number of switching between fixation and saccades and hence *Normalized Fixation Switching* would be higher for easy content. For a given content, the local features capture the variation in the reading pattern with change in time. This variation is measured using the *Entropy* of the local features, which can capture the individual characteristics related to consumption of the content.

**Normalized Fixation Duration**. While reading the content, the eye gaze undergoes multiple fixations with intermittent transition to saccades. The duration of the fixation in every line is normalized with the total duration in reading that line. This is termed as normalized *fixation duration* ($f_l^d$) for the $l$th line of a given content. The aggregate (mean) of all these line features are termed as global feature for *normalized fixation duration*, $Gf^d$ as given in (1), where N is the number of lines in the content.

$$Gf^d = \frac{\sum_{i=1}^{N} f_l^d}{N} \quad (1)$$

**Normalized Fixation Switching**. The switching between the fixation and saccades is measured by the number of transitions between the two. This transition value is derived for every line and normalized with the duration of the line. This is termed as normalized *fixation switching* ($f_l^s$) for the $l$th line of a given content. The aggregate (mean) of all these line features are termed as global feature for *normalized fixation switching*, $Gf^s$ as given in (2), where N is the number of lines in the content.

$$Gf^s = \frac{\sum_{i=1}^{N} f_l^s}{N} \quad (2)$$

**Entropy**. The entropy is derived to quantify the randomness in the features in the behavior of eye gaze during a silent reading task. Initially the local features namely, normalized *fixation duration* and normalized *fixation switching* features are derived for each line. The line features, $f_l \in \{f_l^d, f_l^s\}$ for all the lines corresponding to all the "easy" and "difficult" contents are first extracted separately. This is done for each of the participants. Then, the histograms $\left(h_b \in \{h_b^d, h_b^s\}, 1 \le b \le M\right)$ of the features are derived in M bins, for each type of the content, for every participant. The individual histograms are then converted to probability distribution $\left(p_b \in \{p_b^d, p_b^s\}, 1 \le b \le M\right)$ by normalizing such that the sum of the histogram values for all the bins to 1 as given in (3). Thus the probability distribution of the features is obtained for both the contents. This is done separately for each participant. Here, the $p_b^d$ is obtained from $f_l^d$ and $p_b^s$ is obtained from $f_l^s$.

$$p_b^d = \frac{h_b^d}{\sum_{b=1}^{M} h_b^d}, \quad p_b^s = \frac{h_b^s}{\sum_{b=1}^{M} h_b^s} \tag{3}$$

Next, the Shannon entropies, $H(f^d)$ and $H(f^s)$, are derived for a line features $f_l^d$ and $f_l^s$, respectively, as given by (4). These entropies give insights into the local variations of the fixation durations and frequency of the transition between the fixations and saccades.

$$H(f^d) = -\sum_{b=1}^{M} p_b^d \log_2(p_b^d), \; H(f^s) = -\sum_{b=1}^{M} p_b^s \log_2(p_b^s) \tag{4}$$

## 4 Experimental Results

The analysis is performed to quantify the difficulty level of a content with respect to another using the eye gaze data of readers. The experiments are performed using two sets of stimuli as explained in Sect. 2. There are nine participants where each one is asked to silently read three easy and three difficult texts. The eye-tracking data is collected during the reading. Initially the eye-tracking data is preprocessed where the eye blink regions are detected where the eye gaze data is not available and interpolated linearly using the adjacent data in time. After that the eye gaze data corresponding to each line is derived. Local features namely "*Fixation duration*" and "*Fixation Switching*" are derived for every line.

For each participants the aggregate (mean) of the feature values for all the lines and for all the three texts for each type (easy and difficult) are derived. This provides the overall information on the variation among each participant for a particular type of text. Thus this provides a global information about the content. Figure 6 provides a comparison of these two global features ($Gf^d$ and $Gf^s$) between the two types of texts for all the participants.

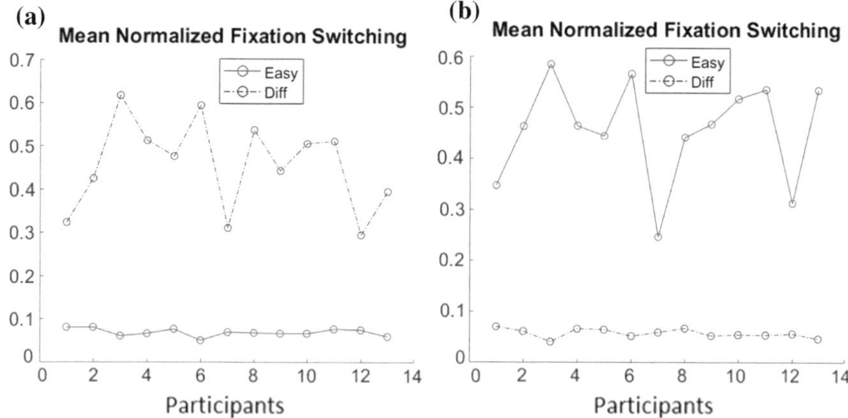


**Fig. 6** Subject wise fixation related global information—**a** Fixation duration normalized with total duration ($Gf^d$), **b** number of switching between fixation and saccade, normalized with total duration ($Gf^s$)

It can be seen that the *mean normalized fixation duration* is less for easy text compared to the difficult text which matches with our intuition. This is coupled with the observation that the normalized number of switching between fixation and the saccades is higher in the easy text that the difficult one. This is due to the fact that for easy text one tends to read fast so the movement from one word or phrase to the next is fast, indicating momentary short duration fixations. Hence the switching is more but the relative fixation duration is less for easy texts. On the other hand for difficult texts, due to the presence of unfamiliar words and difficult to understand phrases, the duration per fixation tends to be more and the switching between fixation and saccades per unit time is less. Thus these global features are clearly able to separate the two types of contents in terms of their difficulty levels.

Another important observation to be made on the variation of the features among the participants. For *mean normalized fixation duration*, the variation for easy text is much less compared to the difficult text indicating that almost all participants are fixating for similar duration. However, in case of *mean normalized fixation switching*, the scenario is reversed, where almost for all participants the amount of switching is similar for difficult texts and varies a lot for easy texts. The variation in the switching for easy text is due to the individual capability of reading speed governed by the efficiency in the movement of the eye. On the other hand the variation in the fixation duration among the subjects for difficult text is mostly governed by the language and vocabulary skill. These variations motivate us to investigate the details within the local features by computing the entropies associated with the individual participants for a given type of content.

To understand homogeneity of the effect of a content on a participant, initially the probability values are derived using the histogram analysis of the local features by partitioning into 10 bins. Then the entropies are derived for each probability distribution obtained from a type (easy or difficult) of content for each participant. The variations in individual levels of entropies $H(f^d)$ and $H(f^s)$ derived from the local features are shown in Table 3 and Table 4 respectively. Higher values of entropy indicates that there is a high amount of variation of a local feature in a given content for an individual, which is the reflection of momentary struggle while reading as against a continuous and smooth flow in reading. From Table 3, it can be seen that participants P2 and P7 are having a large difference in the $H(f^d)$ between the easy and difficult contents with higher values in easy content. This indicates that there is very less variation in the local features for the difficult content and the participants struggled to read throughout the content. On the other hand for P10–P13 it is reverse indicating that they are quite comfortable with the difficult content. These participants regularly read novels as given in the feedback. Similar observations related to the entropies for *fixation switching* can also be seen from Table 4.

## 5 Conclusion

Analysis of the readability of a content is important for the popularity of the content. Standard methods of such evaluation involve NLP-based analysis and statistical parameters related to number of words per sentence, percentage of complex words, average length sentences, etc. However, the information derived from the eye gaze provides direct insights into how an individual is reading the content. We have used a low cost, affordable eye tracking device capture the eye gaze data and derive local and global featured related to fixations and saccades. The time series of the eye gaze data (X–Y coordinates) is segmented to extract the information for each line. Then the fixation duration and the number of times the transition from fixation to saccades are derived for every line, which is then normalized with the line duration to derive the local features. The aggregation of these features over all the contents of same type (easy or difficult) is used to derive the global features, which are useful to indicate the difference in the readability of two types of contents. The local features provide the instantaneous variations while reading a type of content. Thus the entropy derived from the local features can provide individual participants' reading characteristics in finer details. In future, we would like to extend this work to understand the effect of the intermixing of easy and difficult content and also analyze other physiological features including the brain signals. Moreover, we would also like to extend the same to multiple levels of difficulty beyond two levels.

**Acknowledgements** The authors would like to thank the participants for their cooperation during the experiment and providing the data for the analysis. The data is anonymized for the processing and informed consent was taken from the participants. The data collection protocol followed Helsinki Human Research guidelines (https://www.helsinki.fi/en/research/research-environment/research-ethics).

**Table 3** Comparison of the entropies, $H(f^d)$ between the participants for easy and difficult content

| Content type | P1 | P2 | P3 | P4 | P5 | P6 | P7 | P8 | P9 | P10 | P11 | P12 | P13 |
|---|---|---|---|---|---|---|---|---|---|---|---|---|---|
| Easy | 2.12 | 2.25 | 2.16 | 1.97 | 2.22 | 2.0 | 2.41 | 2.18 | 1.70 | 1.78 | 1.72 | 1.78 | 2.26 |
| Difficult | 2.08 | 1.97 | 2.16 | 1.78 | 2.26 | 2.0 | 1.53 | 1.93 | 1.67 | 2.13 | 1.95 | 2.15 | 2.55 |

**Table 4** Comparison of the entropies, $H(f^s)$ between the participants for easy and difficult content

| Content type | P1 | P2 | P3 | P4 | P5 | P6 | P7 | P8 | P9 | P10 | P11 | P12 | P13 |
|---|---|---|---|---|---|---|---|---|---|---|---|---|---|
| Easy | 2.04 | 2.45 | 1.74 | 2.06 | 1.80 | 2.24 | 2.20 | 2.42 | 1.42 | 2.04 | 1.78 | 1.80 | 2.29 |
| Difficult | 2.41 | 2.26 | 2.02 | 2.24 | 1.64 | 2.16 | 1.95 | 2.41 | 2.50 | 2.28 | 2.14 | 1.81 | 2.49 |

## References


1. Mehlenbacher, B., Bennett, L., Bird, T., Ivey, M., Lucas, J., Morton, J., Whitman, L.: Usable e-learning: a conceptual model for evaluation and design. In: Proceedings of HCI International, 22 July 2005, vol. 2005, p. 11
2. Kincaid, J.P., Fishburne Jr., R.P., Rogers, R.L., Chissom, B.S.: Derivation of new readability formulas (automated readability index, fog count and flesch reading ease formula) for navy enlisted personnel. No. RBR-8-75. Naval Technical Training Command Millington TN Research Branch (1975)
3. Stemler, Steve: An overview of content analysis. Pract. Assess. Res. Eval. **7**(17), 137–146 (2001)
4. Goldberg, J., Helfman, J.: Eye tracking for visualization evaluation: reading values on linear versus radial graphs. Inf. Vis. **10**(3), 182–195 (2011)
5. Campbell, Jennifer D., Tesser, Abraham: Motivational interpretations of hindsight bias: an individual difference analysis. J. Pers. **51**(4), 605–620 (1983)
6. Mayer, R.E. (ed.): The Cambridge Handbook of Multimedia Learning. Cambridge University Press, Cambridge (2005)
7. Sweller, John: Cognitive load during problem solving: effects on learning. Cogn. Sci. **12**(2), 257–285 (1988)
8. Gavas, R., Das, R., Das, P., Chatterjee, D., Sinha, A.: Inactivestate recognition from eeg signals and its application in cognitive load computation. In: 2016 IEEE International Conference on Systems, Man, and Cybernetics (SMC), pp. 003 606–003 611. IEEE (2016)
9. Shi, Y., Ruiz, N., Taib, R., Choi, E., Chen, F.: Galvanic skin response (GSR) as an index of cognitive load. In: CHI'07 Extended Abstracts on Human Factors in Computing Systems, pp. 2651–2656. ACM (2007)
10. Ryu, K., Myung, R.: Evaluation of mental workload with a combined measure based on physiological indices during a dual task of tracking and mental arithmetic. Int. J. Ind. Ergon. **35**(11), 991–1009 (2005)
11. Zhai, J., Barreto, A.: Stress detection in computer users based on digital signal processing of noninvasive physiological variables. In: 28th Annual International Conference of the IEEE Engineering in Medicine and Biology Society, 2006. EMBS'06, pp. 1355–1358. IEEE (2006)
12. Nathan, D., Jeutter, D.: Exploring the effects of cognitive load on muscle activation during functional upper extremity tasks. In: 25th Southern Biomedical Engineering Conference 2009, 15–17 May 2009, Miami, Florida, USA, pp. 17–18. Springer (2009)
13. Hjemdahl, Paul, et al.: Differentiated sympathetic activation during mental stress evoked by the Stroop test. Acta Physiol. Scand. Suppl. **527**, 25–29 (1983)
14. Just, Marcel A., Carpenter, Patricia A.: A theory of reading: from eye fixations to comprehension. Psychol. Rev. **87**(4), 329 (1980)
15. Raney, G.E., Campbell, S.J., Bovee, J.C.: Using eye movements to evaluate the cognitive processes involved in text comprehension. J. Vis. Exp. JoVE (83) (2014)
16. Navarro, O., Molina, A.I., Lacruz, M., Ortega, M.: Evaluation of multimedia educational materials using eye tracking. Procedia-Soc. Behav. Sci. **197**, 2236–2243 (2015)
17. Johnson, J.: Designing with the Mind in Mind: Simple Guide to Understanding User Interface Design Rules. Morgan Kaufmann Publishers/Elsevier, Amsterdam, Boston (2010)
18. Burton, R., Saunders, L.J., Crabb, D.P.; Areas of the visual field important during reading in patients with glaucoma. Jpn. J. Ophthalmol. **59**(2), 94±102 (2015). Epub 2014/12/30. https://doi.org/10.1007/s10384-014-0359-8. PMID: 25539625
19. Murata, N., Miyamoto, D., Togano, T., Fukuchi, T.: Evaluating Silent reading performance with an eye tracking system in patients with glaucoma. PLoS ONE **12**(1), e0170230 (2017)
20. Forssman, L., Ashorn, P., Ashorn, U., Maleta, K., Matchado, A., Kortekangas, E., Leppänen, J.M.: Eye-tracking-based assessment of cognitive function in low-resource settings. Arch. Dis. Child. **102**(4), 301–302 (2016)
21. Kliegl, R., Laubrock, J.: Eye-movement tracking during reading, chapter 4. In: Research Methods in Psycholinguistics and the Neurobiology of Language: A Practical Guide, vol. 68 (2017)

22. Johansen, S.A., San Agustin, J., Skovsgaard, H., Hansen, J.P., Tall, M.: Low cost vs. high-end eye tracking for usability testing. In: CHI'11 Extended Abstracts on Human Factors in Computing Systems, pp. 1177–1182. ACM (2011)
23. Salvucci, D.D., Goldberg, J.H.: Identifying fixations and saccades in eye-tracking protocols. In: Proceedings of the 2000 Symposium on Eye Tracking Research and Applications, pp. 71–78. ACM (2000)
24. Popelka, Stanislav, Stachoň, Zdeněk, Šašinka, Čeněk, Doležalová, Jitka: EyeTribe tracker data accuracy evaluation and its interconnection with hypothesis software for cartographic purposes. Comput. Intell. Neurosci. **2016**, 20 (2016)
25. Korfiatis, Nikolaos, GarcíA-Bariocanal, Elena, Sánchez-Alonso, Salvador: Evaluating content quality and helpfulness of online product reviews: the interplay of review helpful-ness vs. review content. Electron. Commer. Res. Appl. **11**(3), 205–217 (2012)
26. Hinkin, Timothy R.: A brief tutorial on the development of measures for use in survey questionnaires. Organ. Res. Methods **1**(1), 104–121 (1998)
27. Hornof, A.J., Halverson, T.: Cleaning up systematic error in eyetracking data by using required fixation locations. Behav. Res. Methods Instrum. Comput. **34**(4), 592–604 (2002)
28. Carpenter, R.H.S.: Movements of the Eyes, 2nd edn. Pion, London (1988)
29. Noton, D., Stark, L.: Scanpaths in saccadic eye movements while viewing and recognizing patterns. Vis. Res. **11**, 929–942 (1970)
30. Senders, J.W., Fisher, D.F., Monty, R.A. (Eds.): Eye Movements and the Higher Psychological Functions, vol. 26. Routledge (2017)
31. Taylor, I., Taylor, M.M.: The Psychology of Reading. Academic Press (2013)
32. Johnston, J.: A filter family designed for use in quadrature mirror filter banks. In: IEEE International Conference on Acoustics, Speech, and Signal Processing, ICASSP' 80, vol. 5, pp. 291–294. IEEE (1980)
33. Jacobson, A.L.: Auto-threshold peak detection in physiological signals. In: Proceedings of the 23rd Annual International Conference of the IEEE Engineering in Medicine and Biology Society, 2001, vol. 3, pp. 2194–2195. IEEE (2001)

# Part III
# Signal Processing and Analytics—II

# Multi-node Approach for Map Data Processing

**Vít Ptošek and Kateřina Slaninová**

**Abstract** OpenStreetMap (OSM) is a popular collaborative open-source project that offers free editable map across the whole world. However, this data often needs a further on-purpose processing to become the utmost valuable information to work with. That is why the main motivation of this paper is to propose a design for big data processing along with data mining leading to the obtaining of statistics with a focus on the detail of a traffic data as a result in order to create graphs representing a road network. To ensure our High-Performance Computing (HPC) platform routing algorithms work correctly, it is absolutely essential to prepare OSM data to be useful and applicable for above-mentioned graph, and to store this persistent data in both spatial database and HDF5 format.

**Keywords** OpenStreetMap · Road network quality · Big data parsing Multi-node processing · ETL · State machine · Pipeline

## 1 Introduction

One of the biggest advantages of OSM [1] is such that everyone can contribute [2]. There is no doubt this factor has helped the project to be successful to the possible extent [3]. Nevertheless, this feature tends to produce a lot of misleading if not missing data. Our objective is to present an approach for processing such a big spatial data to be of better quality.

The main motivation of this paper is encouraged by Antarex [4] project's second use case—a self-adaptive server-side navigation system to be used in smart cities. Such a navigation system surely needs to be backed by a reliable graph-like data for

V. Ptošek (✉) · K. Slaninová
IT4Innovations National Supercomputing Center, VŠB - Technical University of Ostrava, Ostrava, Czech Republic
e-mail: vit.ptosek@vsb.cz

K. Slaninová
e-mail: katerina.slaninova@vsb.cz



R. Chaki et al. (eds.), *Advanced Computing and Systems for Security*, Advances in Intelligent Systems and Computing 897,
https://doi.org/10.1007/978-981-13-3250-0_7

a static routing as well as derived information helping with decisions and planning in a dynamic routing. The more data we are able to process and benefit from, the better we can successfully reach this goal and provide such a service. Our interest is to take advantage of large datasets, as we aim to pay attention to detail on the complete road network, which inevitably brings big data-related tasks and touches usage of parallel algorithms.

This whole data-driven process is supposed to be easily configurable, reusable and efficient. That has been achieved by two mutually communicating services working as process chains as seen in Fig. 1 below.

The first one is meant to obtain and decompress input raw data. Then, we need to filter out desired preprocessed output which from our point of view makes sense to keep as an input for the very next process called parsing and post-processing. In the following step, a job being part of a pipeline is enqueued and kept track of. At this point, a chosen custom metrics are calculated, verified and stored.

Having defined a criterion for a road map data assessment, we have to complete graph with respect to information we need at the final stage of processing. As a result, the network in question can serve for example the routing purposes and traffic data overview.

The rest of the paper is organized as follows. The relevant work in Sect. 2. Exploited data is described in Sect. 3 for better understanding of design clarified

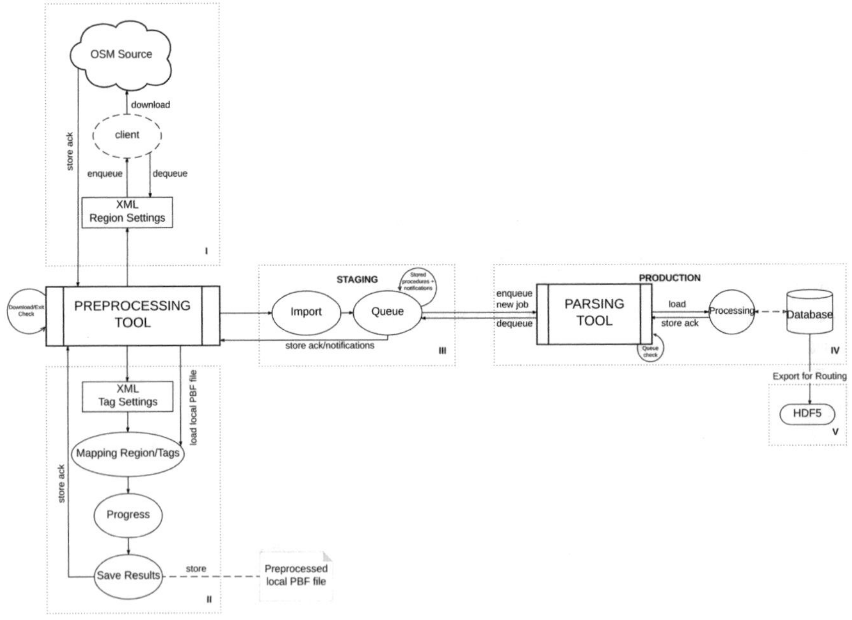


**Fig. 1** Processing design scheme

in Sect. 4 and ETL[1] logic in Sects. 4.1, 4.2 and 4.3. Results and comparisons are explained in Sect. 5. Conclusions and future work are to be found in Sect. 6.

## 2 Related Work

Some research has been done in the field of assessing OSM data quality to show how much trusted such can be [5]. There is couple of methods applicable. Some of them use ground-truth data, determining semantic data quality by co-occurring tags [6]. Another interesting approach rests on machine learning. In that case, unlike ground-truth data, geometrical (length, linearity) and topological (connectivity) characteristics based on segment collocation [7] are highlighted.

Either one proves that there really is inconsistency of tags. Also incorrect or even missing keys (described further in Sect. 3.1) are to be found in original OSM datasets. Looking into our samples, we can confirm that way elements may have little or no information related to road network as well as they might be suspiciously wrong. In spite of this fact, we have decided that there is strong need to complete and normalize this data. Once this is done and validated, we try to calculate the rest with maximal possible accuracy and minimal risk. That being said we came with domain-specific rule sets to be applied. Those are combined with our beforehand approved results that turned out to be correct and reliable. We use semantics to detect related tags and its alternatives if such exist. Geometrical characteristics like number of nodes can give us an idea of road classification. Containing any value we normalize it (unit, type, access, formatting, language), in the opposite we flag them as needing calculations and do so by implication depending on known facts and our previous observations.

Another topic related to this paper introduces an idea of incremental processing chain [8] using VGI[2] data and Osmosis [9]. The described workflow finishes initial processing of whole world OSM data (by then of size as Europe in present) in 8 h, whereas the update takes around 3 h daily due to search operations. In our case, a custom versioning was designed to avoid these computations. Computing itself was desired to focus on high-performance workload to be able to complete several times a day for each country alone.

Finally, a similarly targeted publication discusses routing with OSM data and dealing with its preprocessing [10] for shortest path both online and offline using different routing profiles. Albeit this paper offers a valuable contribution, our intention is to only concentrate on server-side routing using computing power to bring some extra navigation use cases, like emergency or public service routing. To speed routing on the data side, we can generate graph specialized for given task to minimize unnecessary nodes to visit. This means we can process routing graphs for bicycle and car navigation at the same time, each for different kind of work, and decrease routing cost regardless algorithm for finding optimal path used.

---

[1] Extract, transform, load process.

[2] Volunteered Geographic Information.

# 3 Data

The data to be processed are obtained from Geofabrik [11] download server and contains OSM dataset under Open Database License 1.0. The raw data is organized by regions and refreshed on daily basis.

In spite of the fact that everybody can contribute and propagate map data to the source being used, there is always room for inaccurate and flawed or obsolete information we cannot afford to use. Another thing is duplicity or unnecessary data. There have been some limitations experienced with downloading also.

Due to big amount and quality of above-mentioned data, a reduction based on custom settings and sanity check rules has to be made and applied. Detection of missing or broken data is of high priority as some of them can be (re)calculated for use of ours. On the other hand, the rest may be omitted or skipped and discarded.

## *3.1 Input*

Input data file is compressed in Protocolbuffer Binary Format (PBF) [12], which is Google's data interchange format. This data consists of the following:

- **Elements**—the basic components
  - **Nodes**—point on surface defined by latitude and longitude
  - **Ways**—ordered list between nodes that defines polyline (roads in this case)
  - **Relations**—data structure documenting relation between elements
- **Tags**—key value pair belonging to an data element representing its information.

## *3.2 Output*

There are several outputs on the way to the final one. Both preprocessed and parsed file are stored in binary formats for which PBF and HDF5 [13] have been found suitable. The final results are also stored in spatial database [14] as graph of complete coverage of custom roads, segments and additional metrics and information.

We are only interested in roads that we calculated as accessible by car. For such roads, we gather information like whether it is one way, tollway or bridge, what classification it belongs to, what is its maximal allowed speed, etc. For every road we have 22 tags we are able to parse or calculate values for. We add three more custom tags for our use only that gives us extra information for routing.

# 4 Design

The whole process scheme is divided into several parts (see Fig. 2) and each one of them solves separate set of tasks based on its order, type and logic. Two main parts of a single whole are Preprocessing Tool (Sect. 4.1) and Parser (Sect. 4.2). These are further consisting of subprocesses and subsections. Our idea is to run described architecture as a remote service under various parameters due to scalability as described in Sect. 5.1.

The solution as described had been implemented from scratch for both Windows stack and Linux cluster deployment. At this point of a time it is optimized on several levels and able to run on both platforms with the similar if not same features like shutdown signal handling and shell commands execution, just to name a few. The runtime environment comparison is to be seen in Sect. 5.2.

To fully exploit multi-node architecture, besides of a multi-thread, concurrent and parallel processes, certain changes have been made to provide with optional dynamic load balancing between instances reflecting the cluster PBS [15] job scripts where possible. Also some parts, like database import explained in Sect. 4.3, can be run in the background.

## 4.1 *Preprocessing*

The very beginning of the data processing starts with loading settings for maps and tags we are interested in. The behaviour of a run depends on switch options. Preprocessing tool manages the queue and pipeline as a control element and it is designed as a remote data-pump checking a state of a job. This means that every map we want from download queue goes through an implemented chain starting here and only picks the data we need to be saved. It looks after communication, importing borders and new jobs to start on as well as can be seen in Fig. 1.

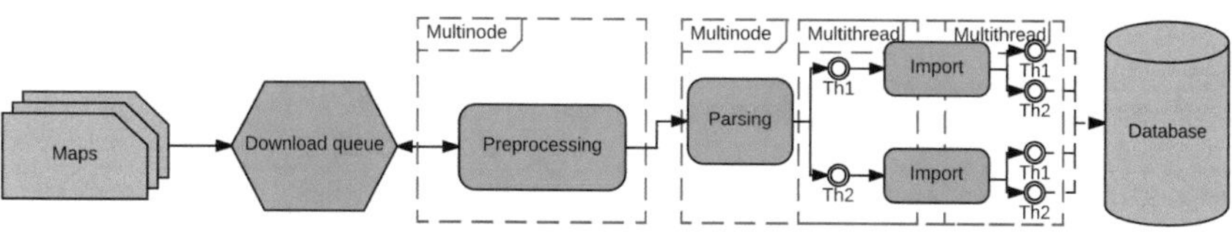


**Fig. 2** Multi-node and parallelism role in scheme

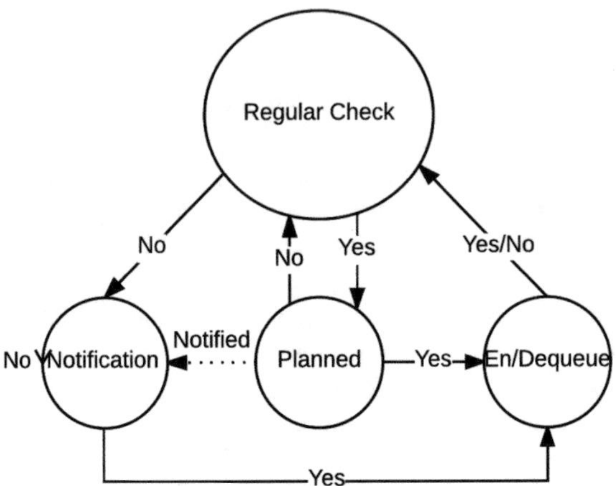


**Fig. 3** Processing represented as a state machine

**State Machine** Preprocessing tool is realized as a state machine (Fig. 3) which allows to run in loops based on a phase it is in. The benefit of this is that we do not need to restart process because of modifications made from the outside as the whole run can be forced remotely on the fly due to the updated pipeline it keeps track of.

**Pipeline** Every area being processed has its own separate pipeline in a queue (see Table 1) saying what process is responsible for a certain planned task and which one is following. There is a given order that has to be followed, although different maps can be processed at the same time, only time-shifted.

The steps of a pipeline are matter of customization and serve to speed-up the process as the first map can be parsed meanwhile the second one is being preprocessed for another parser.

**Table 1** Example of a pipeline queue

| Source | Target | Task | Info | Country | Done |
|---|---|---|---|---|---|
| stored_ function | preprocessing | regular_check | <switch parameters> | | t |
| preprocessing | preprocessing | downloading _map | <map PBF file>-<size>MB | <name> | t |
| preprocessing | preprocessing | extracting | <map PBF file> | <name> | t |
| preprocessing | parsing | parse_pbf | <storage path> | <name> | t |
| preprocessing | preprocessing | import | <boundary filen>>><db> | <name> | t |
| parsing | osm2pgsql | export_country | async | <name> | t |
| parsing | stored_procedure | boundary_cut | staging-asynchronous | <name> | t |

(continued)

**Table 1** (continued)

| Source | Target | Task | Info | Country | Done |
|---|---|---|---|---|---|
| parsing | stored_procedure | deploy_to_production | production-synchronous | <name> | t |
| parsing | stored_procedure | reprocess_ changes | production-synchronous | <name> | t |
| preprocessing | log | exit | <planned/unplanned> | | t |
| preprocessing | log | exit | <cluster node>\|<reason> | | t |
| parsing | log | exit | <cluster node>\|<reason> | | t |

**Planning and Reprocessing** Since the pipeline queue is available for use, it can be taken advantage of in the planning-ahead manner. It also serves its purpose when it comes to abortion of a task, because the process(es) in charge of a cancelled one can easily catch up again and continue from when it was.

Instead of going through the complete pipeline from the very beginning, starting over and redoing what has been done already, a reprocessing point is detected for a new start as illustrated in Fig. 4.

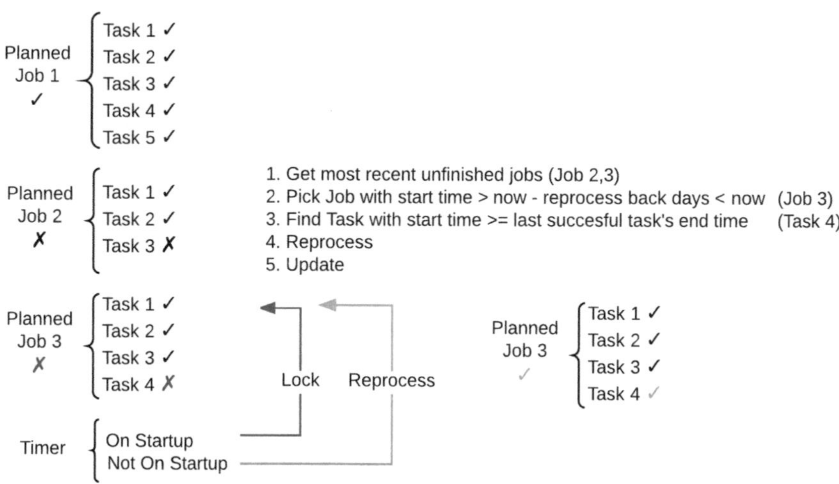


**Fig. 4** Logic of pipeline reprocessing when enabled

## 4.2 Parsing

Parsing instance always follows a leading Preprocessing one and does what it is requested for and lets the Preprocessor know about the progress of a given task. It reminds of master-slave approach based on ETL.

The objective of parsing is to take the data that has not been ruled out (and thus found needful) and to only keep nodes already belonging to the preprocessed roads, which also helps with a compression. The ways are supposed to represent road network for routing. The better information we have about the roads, the more valuable the graph for navigation and visualization would be.

Some data can be extracted from datasets and cleaned afterwards, but most of them can be missing or misleading, so Parser uses our predefined set of rules to (re)calculate metrics and gather basic and additional information of a road and then proceeds to managed indexing of way and node elements of a currently processed country into the road network.

Given as example, that we know some road being currently processed is situated somewhere in the Czech Republic and only comes with tags *lanes:3, oneway:yes*. We can deduct missing classification tag which (now) would be *highway:motorway*. Since we also know there is no restriction, we can lookup an upper speed limit of a mentioned country for a specific type of highway, and thus can also add *maxspeed:130* and most likely *toll:no*, otherwise the toll would be stated.
Parser can run under several settings and in one of two checking modes

1. Continual—on standby waiting for notification scheduling task
2. Periodical—self-scheduling future tasks and waiting for their time to come.

**Queueing** As mentioned in Sect. 4.1, the continuous process leans on a pipeline queue. That brings enqueuing and dequeuing logic into a question. It is also a special case of a log.

In case of more than one instances, it is extremely important to ensure that concurrent ones are not stealing jobs from each other. Not only it would be ineffective to work on the same thing several times, but it could lead to a damaged untrustworthy data. The same stands for the real opposite where some task would not be picked by neither one of the running workers at all. For these cases it is necessary to use locks, double-checking, and sometimes timers in a custom specially designed structures for handling this.

Queueing, forcing start and stop as well as reprocessing and reloading parameters and settings can be done remotely via database commands and notifications.

**Import** Importing processed data is multi-thread asynchronous awaitable process that carries out preparing and moving data into a staging spatial database. Once this is done upon request from pipeline, everything is ready for the next step completely being executed in a database which is a phase of preparing data for a production.

### 4.3 Database Layer

Stating an obvious, a fair amount of work is done within a database layer because of geometry data type values stored in spatial structures. In our case, it also brings advantages that stored procedures give to us. That is why it is so important to pay attention to detail, especially when talking about indexing, prepared statements, query optimization; last but not least, significant communications happen here.

**Spatial Data** It is not only crucial to have information about nodes and ways as they go and are related to each other, but also to be able to represent them in a graphical way—to visualize. For querying geometry data, it is unnecessary to support spatial database so we can link GPS point to a specific road, tell whether some bridge overlaps any highway and many more.

**Automatisation and Asynchronous Awaitable Functions** In regards to database stored procedures and spatial approach, the process running in the DBMS combines both of them. When some country is successfully stored in preliminary tables, there always are some roads crossing the neighbour country border (Fig. 6). The correct way is to only store the part belonging to a specific country before generating the final graph for navigation routing, so those lines are cut on a borderline having the same node ID on both sides down to logically linking them back together on a precisely measured spot.

This is, however, quite costly and there is no need for any application calling this function to be blocked by waiting for a specific result. One way how this is handled is that the stored functions call each other in a specified order. The better one is that the first one called is run asynchronously in the pipeline and the Parser gets back to it once the database gives the result. Meanwhile, it takes another task from the queue available. We can manage to split whatever area we desire in that manner and apply to divide-and-conquer technique.

**Audit and Versioning** Every task is logged on a specific level, but there is one special case when we want to log custom changes so we can take them into account for the next time processing. The bright side of this takes place in incremental reprocessing, history view, no redundancy and taking no risk in overwriting changes with someone else as we prefer ours as long as they are up to date.

When some change is propagated into the downloaded dataset we use, its element version changes and so it is important to us to keep our changesets and versions too. Every country table with the road network has its very own audit table rolled out dynamically and thanks to that we can maintain the most recent and actual versions and also recover from our backups.

# 5 Results

Data processing tool is used at least once a month automatically to mainly catch up with major updates, data enrichment (Fig. 5) and graph readjustment (Fig. 6). In spite of that, it can be run whenever needed.

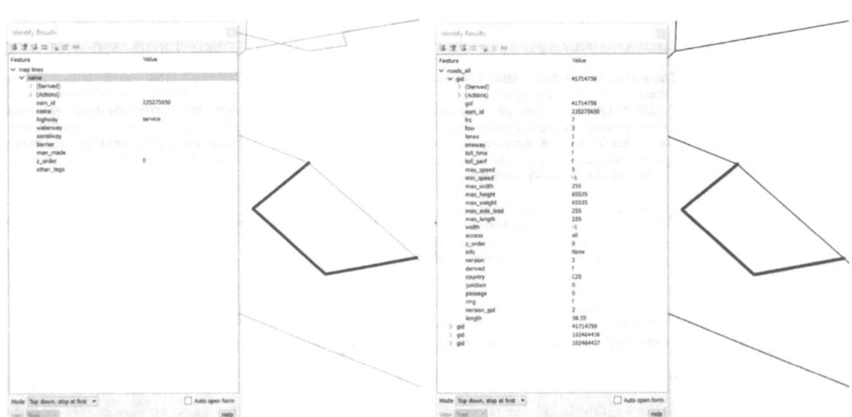

(a) OSM Original with poor tagging (b) OSM Processed with extra information

**Fig. 5** OSM road prior and after processing

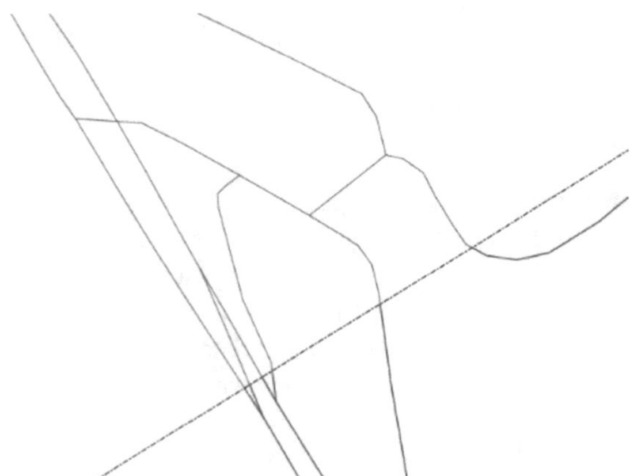

**Fig. 6** Graph split on border

## 5.1 Single-Node Versus Multi-node

It is only needed to have one instance of each part for complete processing in a single mode, but our observations based on measurements showed us to launch three Parsers for one Preprocessor when considering more than one instance. It is very important to take into account the computational efficiency in case of processing difference as well as time shifting—especially in tasks involving big data as large maps are.

To keep all the parts fully utilized, a lightweight resolver has been implemented so the download queue is split between all the active Preprocessors from the map list. Every one of them then tells the first free Parser what to do regardless the node they run on as the pipeline is shared. This helped us to accelerate by « 55% as seen in the Fig. 7.

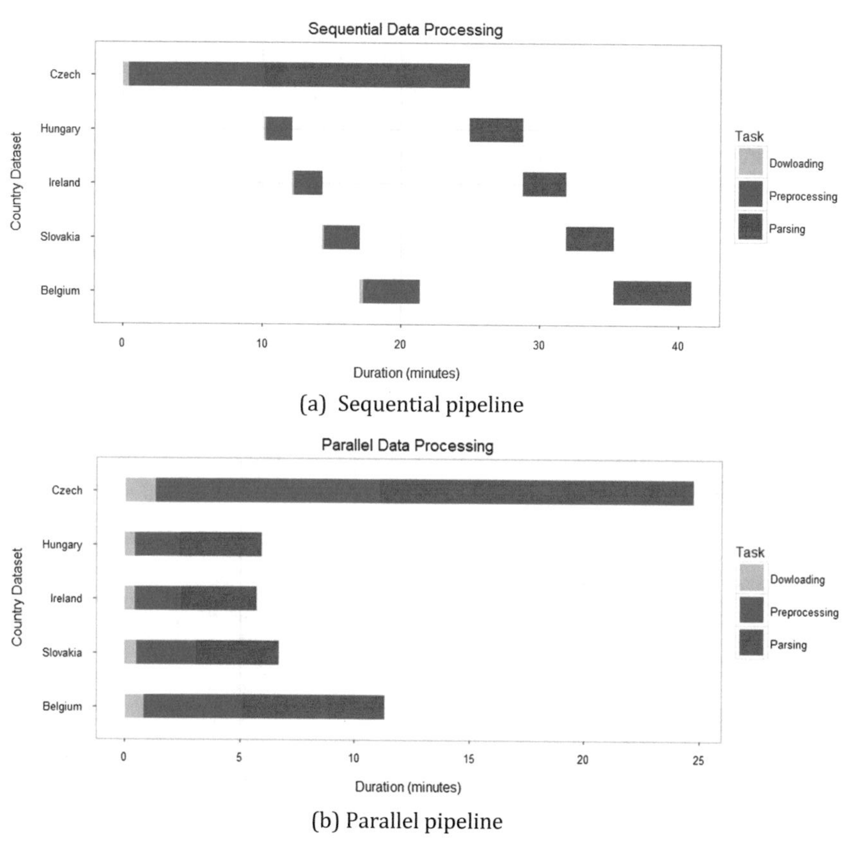


**Fig. 7** Pipeline speed-up

If the processing is found to be short on resources available, it is possible to add up workers on the fly to finish sooner scaling both up and out without interrupting the ongoing job.

## 5.2 Runtime Environment Comparison

There are measured results representing the duration of complete process imaged in Fig. 8. This process consists of scheduled subprocesses and tasks shown previously in Table 1. In this case, a map of France has been chosen for benchmarking as it is the largest available European dataset which makes it suitable because bigger data to process helps us express the diversity of taken execution time better as the difference rise along. The binary file sizes of used map varying during the processing lifetime along with a count of nodes and vertices are exposed in Table 2.

By complete process we mean both preprocessing and parsing part, as seen in Fig. 1. In both cases, the given testing scenario used the same pipeline without the interruption. Although the total duration almost equals, the time ratio of these parts differs obviously. The first part is CPU-bound whilst the latter is memory-bound. We also need to take into account an IO overhead and network speed causing some divergences, especially when talking about virtualization.

These results have been acquired on Salomon [16] supercomputer running single computing node using 24 CPU cores and 128 GB of RAM with disabled hyper-threading. Data has been stored on a hard drive in the Lustre file system.

The aim of this experiment was to prove that the whole process is feasible to run regardless to operation system. The tested tool provides repeatedly exactly the same usable results in very comparable processing times and manner.

These results were also correct in both cases and observations. This means that it is possible to combine different parts on various platforms and run in parallel.

**Table 2** Map set of France during processing

| State of dataset file | PBF size (~GB) | Edges/vertices (~mil) |
|---|---|---|
| Raw downloaded | 3.65 | 5.4/379.2 |
| Parsed roads | 2.83 | 5.4/50.2 |
| Extracted and processed graph | 17.29 | 5.2/50.2 |
| Indexed final graph | 2.03 | 5.0/55.9 |

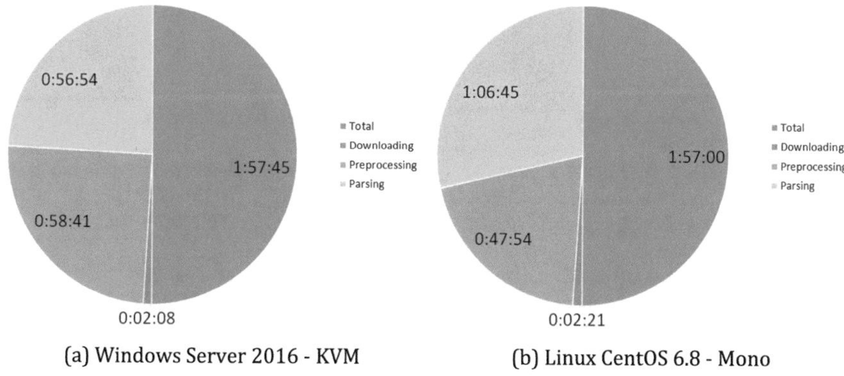


(a) Windows Server 2016 - KVM (b) Linux CentOS 6.8 - Mono

**Fig. 8** Task execution time comparison—same dataset under different environments

## 6 Conclusion and Future Work

As the volume growth of big data for smart cities rises along with its need, the processing time increases also. The way of acquiring applicable information in a feasible time must often be changed. This paper proposes our design for such data processing resulting in an extensive road network routing graph.

The described big data processing itself is in a state that completes a whole map of Europe nowadays. The time of total processing depends on many factors, but under certain settings and adjustments we are confident of a job to be done in 2 h when countries processed simultaneously, with a full result.

We have successfully tested and proved our concept of navigation using this data for the road network and found it working well. Not only it guarantees the parsed road network is suitable for routing, it also shows that values provided by us, like maximal allowed speed computed[3] where not listed by OSM, can be used for prioritizing some paths over another.

An advantage we take of chosen approach is that we can export to different file formats and vice versa. This is especially useful when it comes to querying, visualizing or editing, which is more comfortable via spatial database. On the other hand routing algorithms run faster with use of HDF5 files.

As we can exploit described auditing and map editing[4] into routing systems for our own benefit, an automatic propagation to OSM database would be possible. This can also serve purposes for historical development of road system.

**Acknowledgements** This work has been partially funded by ANTAREX, a project supported by the EU H2020 FET-HPC program under grant 671623, by The Ministry of Education, Youth and Sports of the Czech Republic from the National Programme of Sustainability (NPU II) project 'IT4 Innovations excellence in science—LQ1602'.

[3]For example based on road classification and number of lanes.

[4]That is forcing our changes locally; correcting road information from OSM dataset.

## References

1. OpenStreetMap. https://www.openstreetmap.org
2. Haklay, M., Weber, P.: Openstreetmap: user-generated street maps. IEEE Pervasive Comput. **7**(4), 12–18 (2008)
3. Neis, P., Zielstra, D.: Recent developments and future trends in volunteered geographic information research: the case of openstreetmap. Futur. Internet **6**(1), 76–106 (2014)
4. Autotuning and Adaptivity appRoach for Energy Efficient eXascale HPC Systems. http://www.antarex-project.eu
5. Haklay, M.: How good is volunteered geographical information? a comparative study of openstreetmap and ordnance survey datasets. Environ. Plan. B Plan. Des. **37**(4), 682–703 (2010)
6. Davidovic, N., Mooney, P.: Patterns of tagging in openstreetmap data in urban areas. In: Proceedings of GISRUK (2016)
7. Jilani, M., Corcoran, P., Bertolotto, M.: Automated highway tag assessment of openstreetmap road networks. In: Proceedings of the 22nd ACM SIGSPATIAL International Conference on Advances in Geographic Information Systems, pp. 449–452. ACM (2014)
8. Goetz, M., Lauer, J., Auer, M.: An algorithm based methodology for the creation of a regularly updated global online map derived from volunteered geographic information. In: Proceedings of the Fourth International Conference on Advanced Geographic Information Systems, Applications, and Services, Valencia, Spain, vol. 30, pp. 50–58 (2012)
9. Osmosis. https://wiki.openstreetmap.org/wiki/Osmosis
10. Luxen, D., Vetter, C.: Real-time routing with openstreetmap data. In: Proceedings of the 19th ACM SIGSPATIAL International Conference on Advances in Geographic Information Systems, pp. 513–516. ACM (2011)
11. Geofabrik. https://www.geofabrik.de
12. Protocol Buffers. https://github.com/google/protobuf/
13. The HDF Group. https://www.hdfgroup.org
14. PostgreSQL. https://www.postgresql.org
15. Portable Batch System. https://www.nas.nasa.gov/hecc/support/kb/portable-batch-system-(pbs)-overview_126.html
16. Supercomputer Salomon Hardware Overview. https://docs.it4i.cz/salomon/hardware-overview/

# Naive Bayes and Decision Tree Classifier for Streaming Data Using HBase

**Aradhita Mukherjee, Sudip Mondal, Nabendu Chaki and Sunirmal Khatua**

**Abstract** Classification in real-time environment on streaming data set is one of the most challenging research areas nowadays. Data streaming is used in real-time environment where massive volume of data is generated in small sizes chunks which need to be processed very fast. HBase is a good option which is used for storing such heterogeneous massive small data files in a way so that scalability and availability are preserved. In real-time environment, data are generated exponentially. Thus to store auto incremented data, dynamic splitting is needed which is supported by HBase. We choose tobacco-affected student record and observed that Naive Bayes classifier is less complex and more accurate than decision tree. Also, in real-time environment, it shows its efficacy compared to others when the training sample is too large which is handled by HBase. The key value store in HBase provides the classifiers an extra edge by improving its performance in terms of time.

**Keywords** Big data · HBase · Naive Bayes classifier · Real-time classification
Data streaming · Scalability

---

A. Mukherjee (✉) · S. Mondal · N. Chaki · S. Khatua
Department of Computer Science & Engineering, University of Calcutta,
Kolkata, West Bengal, India
e-mail: aradhita.mukherjee.2016@gmail.com

S. Mondal
e-mail: sudip.wbsu@gmail.com

N. Chaki
e-mail: nabendu@ieee.org

S. Khatua
e-mail: enggnimu_ju@yahoo.com



R. Chaki et al. (eds.), *Advanced Computing and Systems for Security*,
Advances in Intelligent Systems and Computing 897,
https://doi.org/10.1007/978-981-13-3250-0_8

## 1 Introduction

The growth of information technology, sensor technology, healthcare system, online transactions, etc., brings a new era for collecting and analyzing huge volume of variable data with rapid velocity in an efficient way. One of the most challenging areas in big data analytics is the classification in real-time environment on increasing data sets. Traditional rules of classification do not meet the needs of today's requirement; thus improvisation on the existing technology has paid a great attention. In real-time environment, data is generated in the form of a stream, which is boundless, have massive velocity and yet not well-ordered. The application of data stream classification is to predict some instances from the data stream that provides related information about previous instances.

Analytics is used to identify data patterns from the huge amount of valid information by analyzing them, however to recognize the validity of data, real-time data streaming is used. The modern application requires fast classification with higher precision for identifying to which category a new observation is belonging. A Naive Bayes classifier has no complicated iterative parameter estimation which makes it a widely used probabilistic classifier on very large data sets.

Each day, the amount of data and the number of changing data sources continues to grow exponentially, to store such data requires an adequate scaled infrastructure which allows storing, processing, and responding to random access. Web-generated data are generally binary large objects and character large objects; storing them in database make the identification of the actual content and processing those patterns is difficult. This unstructured or structured data requires distributed, fault-tolerant, highly scalable NoSQL database. In this paper, we use HBase to store data of a survey on youth tobacco and apply Naive Bayes classifier to identify to which set of categories an instance belongs. HBase supports random, real-time consistent reads and writes which plays an important role in real-time analysis.

Cancer is a complex group of disease with many possible causes, and tobacco is one of them. The direct and indirect use of tobacco specially smoking damages our heart, i.e., heart attack, coronary heart disease, and blood circulation. Smoking also causes cancer of the lung, tongue, mouth, throat, nose, stomach, liver, kidney, and bone marrow (myeloid leukemia). Smoking causes 84% of deaths from lung cancer and 83% of deaths from Chronic Obstructive Pulmonary Disease (COPD). (American Cancer Society) [1]. So, here, we perform real-time analysis to predict number of affected high school and mid school student affected by tobacco.

The rest of the paper is organized as follows. Section 1 contains problem statement, Sects. 2 and 3 contains related work and background work. Section 4 presents the proposed approach for improving the existing schemes. Section 5 investigates the performance of the proposed scheme through simulation and Sect. 6 presents the conclusion and discusses the future work.

### 1.1 Problem Statement

Data grows exponentially in every sector; such heterogeneous stream data cannot handle by traditional databases. Therefore, we can use the distributed framework like Hadoop, HBase for storing, accessing those data and may use classifier for analyzing those data. HDFS handles large blocks of data efficiently, although it does not show its efficiency when it handles small blocks of data in real time. HBase stores online huge data and allows random read and write operation on them which is not supported by the Hadoop HDFS structure. HBase is a distributed, scalable structure that stores huge data as key/value pairs to enhance scalability.

Classification is usually used to categorize different objects. Here, we used Naive Bayes as it is simple and easy to implement and also suitable for large data set. HBase with classification is applicable in many fields like health care industry, automatic medical diagnosis, etc. Our paper attempts to classify youth tobacco user data into different categories based on how much they are affected by cancer with the help of HBase in real-time environment.

## 2 Related Works

### 2.1 Classifier

This survey paper [2] mainly compares the performance of classification model, which will predict semester grade of final year student for campus placement. The main focus of this paper is to establish a new era of research work that deals with Big Data analytics in educational data mining. They use modified decision tree and rule induction method to evaluate performance. Classification of big data is a great challenge. Categorizing data is the most effective and efficient use of classification. While analyzing large scale, hierarchical learning approaches to analyze patient records will give better results than the other and also computationally more efficient. Enhancing performance of classification algorithm depends on selection of feature set. Highly Correlated Feature Selection Algorithm (HCFS) combined with Hierarchical algorithm is analyzed with the result of the existing Hierarchical algorithm [3].

Based on emotion of lyrics Chinese music is classified. Naive Bayes text classifier is used here, which assume the words of the lyrics are independent and has equal weight. Four different data sets are used for training which shows different accuracy. Naive Bayes is applied on lyrics for emotion detection of music [4].

Naive Bayes classification algorithm provides high classification accuracy on large data set but a slight adjustment in weight-enhancing helps to increase accuracy in the small sample set. Each document is considered as a Poisson random variable generated by the multivariate Poisson model. Combining Poisson distribution model and Naive Bayes shows the accuracy of classification on small sample set [5].

Independence assumptions of attribute restrict classification performance. The method proposed by the author of this paper is based on the estimation of distribution. Here, estimation of distribution algorithms and the Gini index are used as the fitness function to optimize the training set, and then Naive Bayes Classifier provides final classification label [6].

Naive Bayes is used to extract the significant pattern from student's data on real to monitor it at the college level. Naive Bayes shows its efficiency in the classification of student data where data set are generated in real time [7].

Decision tree is mainly used in real-life application because of its interpretability. The author integrates $i^+$ Learning (Intelligent, Incremental, and Interactive Learning) theory to change original decision tree learning algorithms and makes it suitable for real-time analytics. The limitation of this paper is they use binary tree [8].

The novelty of this paper is the implementation of KS-tree algorithm based on decision tree. The limitation of this algorithm is robustness [9]. Improvisation of this paper is the use of J48, which is a version of a decision tree algorithm. Classification model helps to improve accuracy at preprocessing level. Paper states the method for improving the accuracy of decision tree mining with data preprocessing. Supervised filter discretization is used on the data set to increase accuracy [10].

### *2.2 Big Data*

The corelation between characteristics of Big Data (BD) and Healthcare Big Data (HBD) was established here. A conceptual context need diagram in the healthcare sector is used to identify the roles of the entities, which explains the sources that generate data and emphasize the clients who consume each data type. Depending upon who consumes what data, an analytic model was suggested. The main focus of this paper is to bring out the rich characteristics of Healthcare Big Data (HBD). The hierarchical classification approach can be modified by choosing a different ordering of the V's that might apply to some business application. Most research papers are explaining report for classification or analysis of Big Data (BD). A hierarchical classification model is proposed, in which each level in the hierarchy come up with the appropriate representation of Healthcare Big Data (HBD) that is most suitable for analyzing the data. An integrating "representation AND analytic" is used for which efficient mathematical formalism can be used at different levels of the hierarchy [11].

This paper provides a comprehensive analysis of various clustering and classification techniques (Naïve Bayes, J48, BayesNet, Ensemble) used in big data mining. An ensemble model is established in a distributed environment like Hadoop and MapReduce to gain faster result and scaling up huge quantity of data. Paper does not address the problem of concept drift [12].

Data sets are increasing so rapidly that loading them in a single node becomes a bottleneck. In this paper, the author proposed a presplitting tool that helps to increase the scalability of HBase. In these adaptive techniques, rows are used to select files and keys which are already sorted; helps to read data from the column. After sampling, the split region is computed which helps in load balancing. Overhead of this method is high [13].

Real-time read and write operation, random access, column-oriented storage, high scalability, and high availability makes HBase more suitable for massive data storage in the network community. HDFS is suitable for storing large files and MapReduce helps in batch operations. However, it is not convenient for random access to data and not suitable for small-file storage. Instead, HBase provides real-time access to data and is suitable for storing a bunch of small files. The "hot spot" problem of HBase storage is solved in this paper by pre-partitioning and hash design. Insertion of each data record on HBase table would go through a series of operational rules to set up which greatly reduces the performance of HBase real-time read and write data [14].

Real-time analytics on data stream suffers from drift so there is a needs adaptive predictive model. Here, Micro-Cluster Nearest Neighbor (MC-NN) data stream classifier is used, as it works as a parallel classifier without residing data in memory and increase scalability. Each node of a cluster has a mapper which produces output data stream that is aggregated to reduce final output which helps in parallelism. Adaptation and training are achieved by simultaneous distribution of a training instance among nodes in the cluster, which helps them to handle with drift. This algorithm is implemented on Hadoop HDFS framework, which handles large blocks of data efficiently, but cannot handle small, blocks of data in real-time. The proposed architecture does not support random reads and writes which is one of the needs of real-time analysis; to increase scalability, it adapts complex mechanism [15].

Harmony search algorithm is used for data mining in big data. MapReduce architecture is used which supports harmony search agents in grid infrastructure. Harmony search is used for preprocessing of data series before data mining [16].

For classification, mainly Naive Bayes and decision tree are used, as Naive Bayes has higher accuracy and decision tree has interpretability. So in real-time data analytics, we choose both of this classifier.

## 3 Backgrounds

### *3.1 HBase*

HBase is a distributed key value data store model that enables random read, write operation with low latency on huge amounts of structured data. Architecture of HBase has two subparts, HBase master and HBase region server. HBase master includes metadata and region server which handles those data. Automatic sharding of tables with client accessible API makes HBase popular than other NoSql databases.
**Scalability**: HBase supports horizontal partitions and replication, sharding on multiple nodes helps to achieve horizontal scaling, which increase scalability. Another feature to increase scalability is rebalancing of data on several nodes, HBase supports

automatic data rebalancing. Scalability plays an important role in handling loads of client requests. It has two approaches, namely client-based and server-based, both provide static, fixed connections to handle the client request, which reduces scalability. HBase provides fixed connections to a request centralized coordinator to process those request.

**Filter**: Filtering in HBase provides server-side custom filters to reduce information processed by the client. It also helps to decrease network bandwidth but does not shrink server-side I/O. The filter takes as argument a column family, a qualifier, a compare operator, and a comparator. If the column is not found then all columns correspond to that row will be emitted. If the column is found and comparator operator returns true, all the columns of the row will be emitted otherwise the row will not be emitted.

**Data Streaming**: Nowadays, the demand of stream processing is increasing as large volumes of data Processing is not suitable always. Data streaming is important for accessing a large volume of data in real time with a scalable, available and fault-tolerant architecture. Mathematically, data streaming is defined by ordered pair (s, t), where

(i) s is a sequence of tuples and
(ii) t is a sequence of positive real-time intervals.

Data which is generated continuously by lots of data sources and typically in small sizes are known as stream data. This data needs to be processed sequentially and used for analytics like filtering, correlations, and sampling, etc. For streaming data, we need to consider two main tasks, i.e., data processing and storage both are related to each other. Processing plays an important role which decides how to compute that data after stream received and stored with support, ordering, and consistency.

## 3.2 *Classification*

Classification means categorization of data into different class or labels. Based on problem we choose two algorithm.

**Naive Bayes**: Naive Bayes is the simplest technique of classification where feature values are represented in the form of a vector and probabilistic model assign class labels to instances of the problem. In Bayes classifiers, the source of information is training data set, where each feature is independent of other features. This conditional probability model has a vector to represent the features of a problem. It assigns probabilities to each instance for each of n possible outcomes.

This probabilistic model has a problem if features are large, so basing this model on probability tables is infeasible. By using Bayes' theorem, the conditional probability is decayed, which means $x_n$ is the nth attribute of a total n attributes. For categorical attributes, the conditional probability is p $(x_n \mid C_k)$, which is a tuple of $C_k$ classes. $x_n$

divide tuples of $C_k$ classes into the data set. It is mostly used in the field of machine learning and data mining.

**Decision tree**: A decision tree is a supervised learning algorithm that uses tree-like graph or model, where each internal node is a condition on attribute and branch represents possible consequences of decision and leaf node represent class label. The path between internal nodes to leaf is classification rules. Based on the type of target, variable decision tree has two types, namely categorical variable decision tree and continuous variable decision tree.

## 4 Proposed Framework

In this section, we describe a proposed framework for Naive Bayes and decision tree-based classifier for tobacco-affected student's record on the top of HBase. The framework is depicted in Fig. 1. In the proposed framework, the broker plays the role of a middleware that allows the client to access the classifier in a transparent way.

First of all, the broker populates the HBase cluster with a large volume of training data to fine-tune the parameters of the classifiers. The data sets are distributed in Gaussian order which contains the information about users of tobacco in mid and high school students. Since the large volume of training data may contain some irrelevant information, a preprocessing is applied on the stored data in HBase cluster in order to improve the efficiency of the classifier. Once the parameters of the classifiers are fine-tuned, the client sends the analysis request to the broker to analyze test data stored in the cloud.

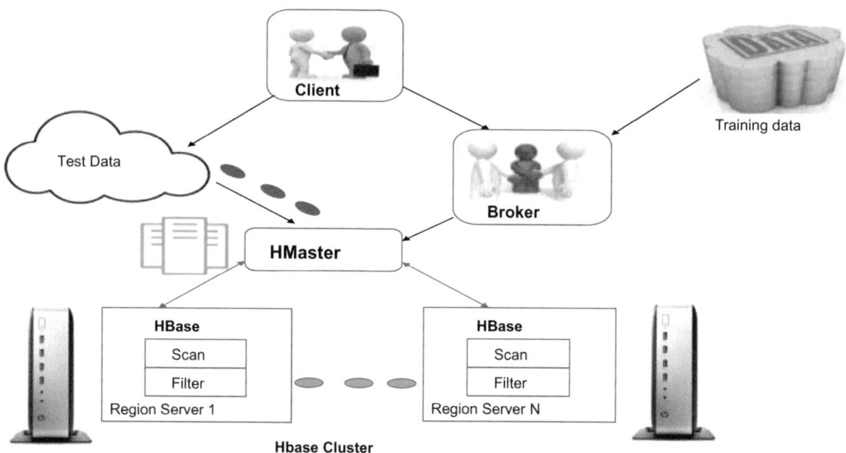


**Fig. 1** Proposed framework

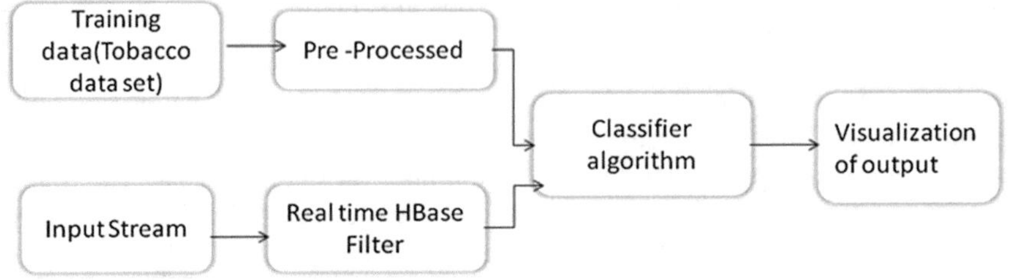


**Fig. 2** Overall workflow

The overall workflow is depicted in Fig. 2 and is summarized by the following steps:

**Algorithm:**

**Step 1:** Training data set are stored in HBase cluster after pre-processing. This training data set is used as the initial population.

**Step2:** Data streaming is applied to the Test Data stored in Cloud in order to make it a real-time data by using micro batch processing.

**Step3:** Data are categorized into different classes by applying classification with the help of HBase filter.

**Step4:** The new populations are then used to estimate the label of test data by Naïve Bayesian and decision tree classifier.

**Step 5:** Step 2 to Step 4 are repeated till the termination criteria is met.

**Step 6:** Finally the confusion matrix is created to compute accuracy and proper visualization of the result.

## 5 Experiments and Result

### *5.1 Data set*

Centers for Disease Control and Prevention (CDC), State Tobacco Activities Tracking and Evaluation (STATE) system have done a survey on the use of tobacco on undergraduate student. YTS collected some features like tobacco smoke, smoking cessation, school curriculum, minors' ability to purchase, or otherwise obtain tobacco products, etc. Data set of tobacco use from the years 2016 to 2017 is available on Data.Gov [1].

### 5.2 *Experimental Setup*

All experiments were performed on machines Intel Core i7 3.60 GHz processor with 4 GB RAM that also contains: CentOS 7 64-bit OS, Apache Hadoop version 2.7.1, and HBase version 1.2.1 which create pseudo-distributed framework (it contains multiple nodes where all the daemons are running on a single node). All the algorithms were implemented in Java: JDK version is "1.8.0_131".

The classifiers decision tree and Naive Bayes applied on stream data sets with sample size range from 3000 to 40,000 records. The average result has been calculated for truthiness of the result after applying classifier on those randomly generated test data. Figure 4 shows that for randomly generated large data set decision tree has the fastest execution time rather than Naive Bayes. Figures 3 and 5 show the comparison between decision tree and Naive Bayes with respect to accuracy and error, the graphs shows with increasing no of data set Naive Bayes shows more accuracy with less error than decision tree. Especially in the case of 10,000 random streams, test data in Fig. 5 shows significant error rate occurred by decision tree, whereas Naive Bayes behave consistently.

## 6 Conclusion

Working with a high volume of data in batch processing system like Hadoop, where data is collected in a batch format usually takes much time to complete the task. One of the major dimensions of Big Data is velocity, so processing speed matters when we apply data analysis. In our method, we have used streaming like microbatch processing, where we add small amount of data streaming to make it in nearly real time, so that it can be classified. We consider here data of tobacco-affected

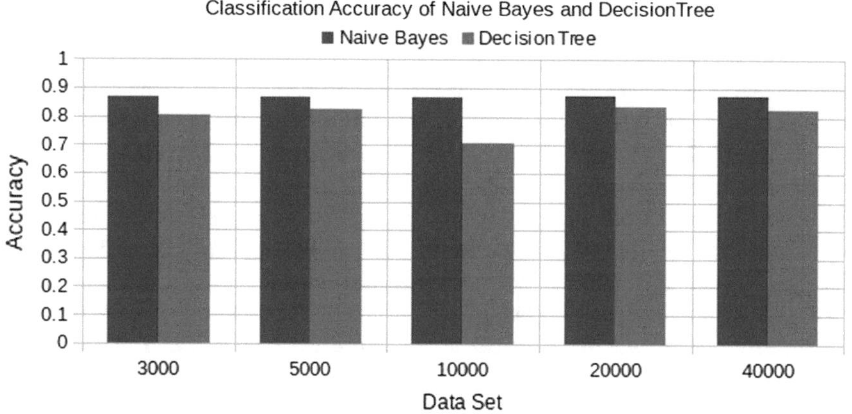


**Fig. 3** Accuracy

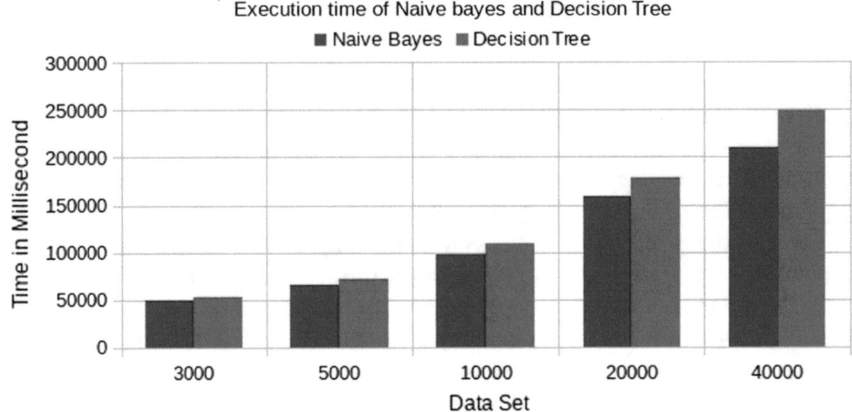


**Fig. 4** Execution time

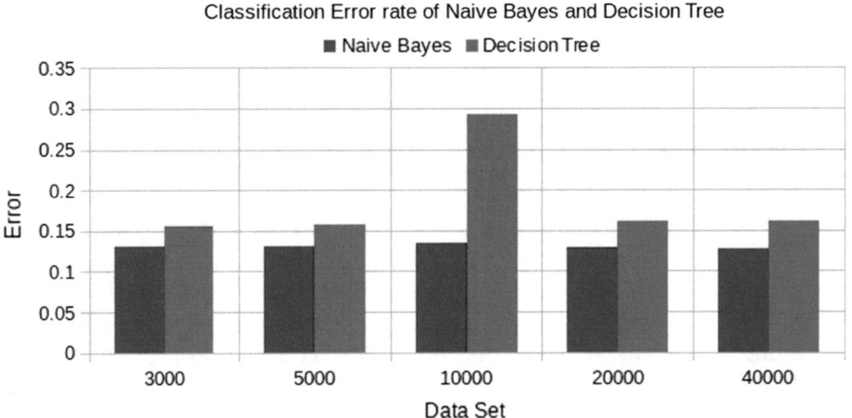


**Fig. 5** Error rate

student and done microbatch processing to stream data, which helps in real-time analytics. In distributed environment, we use HBase filter to perform Naive Bayes and decision tree, and conclude that in real-time environment, Naive Bayes provide greater accuracy and less error than decision tree, but with respect to execution time, it is much faster than others. In the future, we can add more efficient data streaming technique with the help of HBase cluster, Apache Flume, Storm, and Zookeeper. The distribution of data set is considered as Gaussian; by changing it in Poisson distribution may increase the performance of Naive Bayes. Concept drift is a term used to describe changes in the learned structure that occur over time. The occurrence of concept drift leads to a drastic drop in classification accuracy. In our future, we will develop a suitable approach in order to handle concept drift.

## References

1. Guo, J., Xu, W.: Research on optimization of community mass data storage based on HBase. In: Third International Conference on Cyberspace Technology (CCT) (2015)
2. Rajeswari, S., Lawrence R.: Classification model to predict the learners. In: Academic Performance using Big Data. 978-1-4673-8437-7/16/$31.00. IEEE (2016)
3. Vinod, D.F., Vasudevan, V.: A filter based feature set selection approach for big data classification of patient records. In: International Conference on Electrical, Electronics, and Optimization Techniques (ICEEOT) (2016)
4. An, Y., Sun, S., Wang, S.: Naive bayes classifiers for music emotion classification based on lyrics. In: IEEE/ACIS 16th International Conference on Computer and Information Science (ICIS) (2017)
5. Huang, Y., Li, L.: Naive bayes classification algorithm based on small sample. In: IEEE International Conference on Cloud Computing and Intelligence Systems (2011)
6. Yang, X., Dong, H., Zhang, H.: Naive bayes based on estimation of distribution algorithms for classification. In: First International Conference on Information Science and Engineering (2009)
7. Tennant, M., Stahl, F., Rana, O., Gomes, J.B.: Scalable real-time classification of data streams with concept drift. Futur. Gener. Comput. Syst. **75** (2017)
8. Balicki, J., Dryja, P., Korłub, W.: Harmony search for data mining with big data. In: Saeed, K., Homenda, W. (eds.) Computer Information Systems and Industrial Management. CISIM 2016. Lecture Notes in Computer Science, vol. 9842. Springer (2016)
9. Samchao, F.: An incremental decision tree learning methodology regarding attributes in medical data mining. In: IEEE Proceedings of the Eighth International Conference on Machine Learning and Cybernetics, Baoding (2009)
10. Chen, J., Wang, T., Abbey R., Pingeno, J.: A distributed decision tree algorithm and its implementation on big data platforms. In: IEEE Data Science and Advanced Analytics (DSAA) (2016)
11. Chandrasekar, P., Qian, K., Shahriar, H., Bhattacharya, P.: Improving the prediction accuracy of decision tree mining with data preprocessing. In: IEEE Annual Computer Software and Applications Conference (2017)
12. Wan, K.Y., Alagar, V.: Characteristics and classification of big data in health care sector. In: International Conference on Natural Computation, Fuzzy Systems and Knowledge Discovery (ICNC-FSKD) (2016)
13. Gandhi Bhagyashri, S., Deshpande Leena, A.: The survey on approaches to efficient clustering and classification analysis of big data. In: International Conference on Computing Communication Control and Automation (ICCUBEA) (2016)

14. Azqueta-Alzúaz, A., Brondino, I., Patiño-Martinez, M., Jimenez-Peris, R.: Massive data load on distributed database systems over HBase. In: 17th IEEE/ACM International Symposium on Cluster, Cloud and Grid Computing (CCGRID) (2017)
15. https://catalog.data.gov/dataset/youth-tobacco-survey-yts-data
16. Dangi, A., Srivastava, S.: Educational data classification using selective Naïve Bayes for quota categorization. In: 2014 IEEE International Conference on MOOC, Innovation and Technology in Education (MITE), Patiala, 2014, pp. 118–121

# FPGA-Based Novel Speech Enhancement System Using Microphone Activity Detector

**Tanmay Biswas, Shuvadeep Bhattacharjee, Sudhindu Bikash Mandal, Debasri Saha and Amlan Chakrabarti**

**Abstract** In this paper, we have proposed field-programmable gate array (FPGA) based design and implementation of a novel speech enhancement system, which can work for a single microphone device as well as that of a dual microphone device providing background noise immunity. We proposed a microphone activity detector (MAD), which detects the presence of single or dual microphone scenario. After detecting the microphones, multiband spectral subtraction technique enhances the speech signal from different background noisy surrounds. We have implemented our proposed design in Spartan 6 LX45 FPGA using Xilinx system generator tools. The evaluation of the quality of speech of enhanced signal and its correctness of MAD to detect the single or dual microphone system implies that our proposed hardware can work as a proper embedded component for hardware-based execution for speech enhancement.

**Keywords** Time delay estimation · Speech enhancement
Field-programmable gate array · Digital signal processor
Multiband spectral subtraction

---

T. Biswas (✉) · S. Bhattacharjee · S. B. Mandal · D. Saha · A. Chakrabarti
A. K. Choudhury School of Information Technology, University of Calcutta,
Kolkata 700098, India
e-mail: tanmay123g@gmail.com

S. Bhattacharjee
e-mail: sbleo1811@gmail.com

S. B. Mandal
e-mail: sudhindu.mandal@gmail.com

D. Saha
e-mail: debasri_cu@yahoo.co.in

A. Chakrabarti
e-mail: acakcs@caluniv.ac.in



R. Chaki et al. (eds.), *Advanced Computing and Systems for Security*,
Advances in Intelligent Systems and Computing 897,
https://doi.org/10.1007/978-981-13-3250-0_9

# 1 Introduction

The objective of speech enhancement is reduction of noise in the speech signal, which has occurred due to a noisy environment. Spectral subtraction method was invented by Boll [1]. In the spectral subtraction method, noise is estimated from the magnitude spectrum of the noisy signal and estimated noise subtracted from the noisy signal where the phase spectrum of the signal remains unchanged. Spectral subtraction for speech enhancement was introduced by Zhang et al. [2], where the subtraction process is done on both real and imaginary parts of the signal. In the real world, noise involves the signal at different time intervals, which is commonly named as colored noise. In [3], the authors have proposed the multiband spectral subtraction technique to reduce colored noise where the spectrum of the signal is separated into several nonuniform frequency bands, and the estimated noise is subtracted from each of the frequency bands. Speech enhancement based on hardware software codesign using FPGA platform can be found in [4, 5]. A low-power dual microphone speech enhancement using FPGA was presented in [6]. The authors have used phase-based filtering in the form of time–frequency masking for enhancing speech signal. The main purpose to use this filter is to maintain the spectral structure of the speech sources. In [7], the authors showed speech enhancement algorithm on a dual microphone system using on the function of coherence of the signals. The proposed scheme handles the coherence between the target and the noise signals for noise reduction assuming that the spacing between the microphones is narrow.

In the microphone array system, we need to adjust the time delay between the signals. If the delay between the signals is not properly adjusted, then the mixing of those signals can generate unwanted noise. Thus, the time ordering of the samples of the signals is an important issue for prevention of the noise. Several algorithms have been used to estimate the time delay between the signals with a deviating degree of accuracy and computational complexity. Find the degree of correlation between the signals, cross correlation [8] method (CC) has been used. By improving the accuracy and computational cost of cross-correlation method is named as generalized cross-correlation methods [9]. By selecting the weighting function in a generalized cross-correlation method, we can converge to the maximum likelihood method [10]. We have preferred the phase transform algorithm for localizing the sound source as it suits the need for real-time applications. In PHAT algorithm, the cross-power spectrum of the signals is normalized to a constant value which provides an identical shrill correlation peak.

From the above discussion, we see that most of the speech enhancement research works has been done either for a dual microphone or that of a single microphone system. In this paper, we propose a novel speech enhancement machine which works for a single as well as a dual microphone system. A microphone activity detector (MAD) detects the presence of single or dual microphone system and then speech signal is enhanced by the multiband spectral subtraction methods [3] from the noisy background environment. Also, if MAD detects the dual microphone scenario, then the time delay between the signals are adjusted before processing of speech enhancement algorithm. By doing this, we can achieve suppression of background noise in single and dual microphone system. The hardware execution can be compiled up in two ways which are the digital signal processors and FPGAs. For the opportunity of dedicated DSP blocks and configurable logic cells involving parallel computing, we have selected FPGA as our objective hardware [11]. The Xilinx system generator platform [12] in MATLAB/SIMULINK environment [13] have been used to design and verify our proposed design. We convey the perceptual evaluation of speech quality (PESQ) measurement of enhanced signals in the dual and single microphone system for different noisy environments. In every condition, the MAD correctly detects the microphones and PESQ score shows significant results in each of the noisy conditions. We have also evaluated the resource utilization, time information, and throughput of Spartan6 Lx45 FPGA for the proposed design. The major contribution of the proposed design is a new approach to detect the microphone (MAD) to enhance the speech signal from background noisy environments of single or dual microphone system and FPGA based implementation of the proposed hardware.

## 2 Background

In this section, we discuss some key consequences associated with the estimation of time delay between the microphones and speech enhancement algorithm to realize our proposed work in the following sections.

### 2.1 *Time Delay Estimation for Two Microphone System*

Time taken by a signal from a source to destination is called time delay. We acquire that dual microphones are received the noisy speech signals $x_1(t)$ and $x_2(t)$ respectively, which can be expressed as

$$x_1(t) = s(t) + n_1(t) \tag{1}$$

$$x_2(t) = s(t + \tau) + n_2(t) \tag{2}$$

where, $s(t)$ is the speech signal and $n_1(t)$, $n_2(t)$ are the noise mixed to the speech signal. Time delay between the signals is $\tau$.

In frequency domain, these signals can be represented as

$$X_1(\omega) = S(\omega) + N_1(\omega) \tag{3}$$

$$X_2(\omega) = S(\omega) + N_2(\omega) \tag{4}$$

The delay between the microphones that maximized the CC between $X_1(w)$ and $X_2(w)$ can be expressed as

$$\tau = argmax \int_{\infty}^{\infty} X_1(\omega)X_2(\omega)e^{-j\omega\beta}d\omega \tag{5}$$

Practically, unfiltered cross-correlation methods does not well perform in the reverberant environment. So, a frequency-dependent weighting function $W(w)$ is applied for minimizing the reverberant effect which can be defined as

$$\tau = argmax \int_{\infty}^{\infty} W(\omega)X_1(\omega)X_2(\omega)e^{-j\omega\beta}d\omega \tag{6}$$

The weighting function based on phased transform can be represented as

$$W(\omega) = \frac{1}{|X_1(\omega)||X_2(\omega)|} \tag{7}$$

The delay between the signals based on phased transform can be mathematically simplified as

$$\tau = argmax \int_{\infty}^{\infty} cos(\angle X_1(\omega) - \angle X_2(\omega) - \omega\beta)d\omega \tag{8}$$

where "$\angle$" define the phase angle between the signals.

## 2.2 Multiband Spectral Subtraction

To enhance the speech signal from the background noisy environment, the spectral subtraction algorithm is mostly used. The fundamental principle of the spectral subtraction technique is noise estimate from the whole magnitude part of the signal, which is subtracted from the noisy magnitude part. In time domain, the noisy signal can be represented as

$$x(t) = s(t) + n(t) \tag{9}$$

where $x(t)$, $s(t)$ and $n(t)$ are the noisy signal, speech signal, and the noisy, respectively, and $t$ is the discrete time index.

In frequency domain, the noisy signal can be represented as

$$X(\omega) = S(\omega) + N(\omega) \tag{10}$$

where $X(\omega)$, $S(\omega)$ and $N(\omega)$ are the frequency domain signals corresponding to $x(t)$, $s(t)$, and $n(t)$, respectively.

To estimate the clean speech $\hat{S}_{\omega}(f)$, estimated noise $\hat{N}_{\omega}(f)$ is subtracted form the magnitude spectrum of the noisy signal $X_{\omega}(f)$ which can be represented as

$$|\hat{S}_{\omega}(f)|^{\gamma} = |X_{\omega}(f)|^{\gamma} - |\alpha \hat{N}_{\omega}(f)|^{\gamma} \tag{11}$$

By averaging the speech samples on silence period, we can estimate the noise spectrum $|\hat{N}_{\omega}(f)|^{\gamma}$ of the signal

$$|\hat{N}_{\omega}(f)|^{\gamma} = 1/L \sum_{i=0}^{L-1} |X_{\omega}(f)|^{\gamma} \tag{12}$$

where the number of samples during silence period of speech is $L$.

The function of signal to noise ratio is over subtraction factor ($\alpha$). In spectral subtraction process, $\alpha$ is subtracted from the whole noisy signal. The magnitude spectrum of the signals is separated into several frequency bands and the subtraction done throughout each of the frequency bands independently to reduced colored noise. The enhanced speech signal of the $i$th frequency bands can be represented as

$$|\hat{S}_i\omega(f)|^{\gamma} = |X_i\omega(f)|^{\gamma} - \alpha_i\delta_i|\hat{N}_i\omega(f)|^{\gamma} \tag{13}$$

where $\delta_i$ and $\alpha_i$ are the tweaking factor and over subtraction factor of each $i$th frequency bands, respectively. Beginning and ending frequency bins of $i$th frequency band are $b_i$ and $e_i$, respectively. The segmental $SNR_i$ for different frequency bands can be represented as

$$SNR_i(db) = 10 * log_{10} \sum_{f=b_i}^{e_i} (|X_i\omega(f)|/|N_i\omega(f)|)^2 \tag{14}$$

$\alpha_i$ can be evaluated using on the $SNR_i$ which is represented as

$$\alpha_i = \begin{cases} 5 & SNR_i < 5 \\ 4 - 3/20(SNR_i) & -5 < SNR_i < 5 \\ 1 & SNR_i > 20 \end{cases} \tag{15}$$

The noise subtraction levels becomes a degree of control in each of the frequency bands using $\alpha_i$. The measurement of $\delta_i$ [3] is expressed as

$$\delta_i = \begin{cases} 1 & f_i < 1KH_z \\ 2.5 & 1KH_z < f_i < FS/2 - 2KH_z \\ 1.5 & f_i > FS/2 - 2KH_z \end{cases} \tag{16}$$

where $FS$ and $f_i$ are the sampling frequency and upper frequency band, respectively.

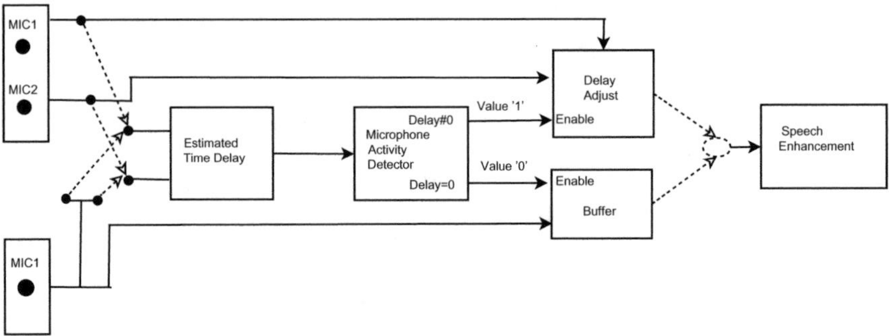


**Fig. 1** Basic block diagram of proposed design

## 3 Proposed Design

We have proposed speech enhancement technique using a microphone activity detector (MAD) to detect the single or dual microphone system. The proposed system consists of two steps: microphone activity detector (MAD) and speech enhancement process. From Eq. (8), we find that if dual microphone signals are the source of the system, then their sample values must be different and $\tau$ holds some value. Using this $\tau$, we need to adjust the time difference between the signals to the proper enhancement of noisy speech signals. If the system source is a single microphone and pass the signal into the dual channel to keep the equivalent paths of dual microphone system, then the $\tau$ become "0" and we do not need to adjust the delay between the signals. So, mathematically, the microphone activity detector (MAD) using Eq. (8) can be represented as

$$\tau = \begin{cases} 1 \; if \; x_1 \neq x_2 \\ 0 \; if \; x_1 = x_2 \end{cases} \tag{17}$$

After processing the MAD and delay adjustment between the signals (if necessary), the noisy signal is processed by the multiband spectral subtraction subsystem as defined by Eq. (14) to enhance the speech signals. The proposed block diagram is shown in Fig. 1.

## 4 Hardware Implementation

Our proposed design has two principle blocks which are microphone activity detector (MAD) and speech enhancement block. MAD measure the time delay between the signals. The time domain noisy signals are converted to the frequency domain by the fast Fourier transform (FFT) hardware which is working in the pipelined version. Cordic Arctan hardware [12] converts these signals in magnitude and phase format. Then, PHAT base time delay is estimated by the accumulator to perform

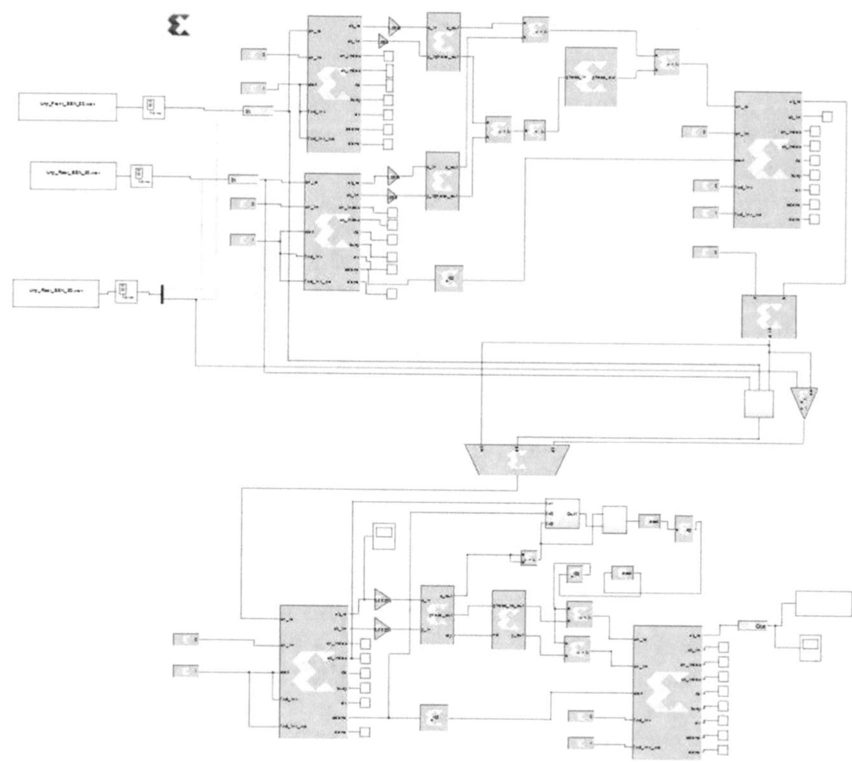

**Fig. 2** Hardware design of proposed system

the continuations addition for scale-based integration. The controller block is used to get the maximum value for the audible sound. After getting the delay, decision block takes a decision that the system is in the single or dual microphone system. The delay adjustment block enables when dual microphone signal processed and when single signal pass then buffers is active. After that, the signal is processed through the speech enhancement block. The synchronized signal is processed by the FFT and CORDIC hardware for frequency domain conversion and magnitude and phase format respectively. Noise estimation block estimates the noise from the magnitude of the signal. A single port hardware is enabled in write mode for the first nine samples when only noise is present, and rest of the sample execute in real mode to estimate the noise. The magnitude spectrum of the signal is separated into the 4 frequency bands to calculate the $\alpha_i$ from each of the frequency bands by the SNR and $\alpha_i$ block. The estimated noise is subtracted from each of the frequency bands with different $\alpha_i$ using subtraction hardware. Enhanced magnitudes are combined using adder block and then enhanced magnitude and phase combined together to process through the IFFT hardware for reconstructing the signals. The proposed system generator hardware view is shown in Fig. 2.

## 5 Experimental Results

Field-programmable gate array holds a reconfigurable logic circuitry in the form of a matrix. High-speed FPGA have fast processing system to execute the major real-time systems. The proposed hardware execution has been performed on Atlys Spartan 6 FPGA board. The sound source of the signals was recording the frequency of 1 kHz by keeping 160 samples as the speech signal and 32,000 samples as the silence period, instead it was generated from the source at various degrees of angle using MS Kinect microphones [14]. The signals were collected at 23 °C with sound velocity as 340 m/s [15] where the Euclidean space between the microphones is 11.475 cm. All of the noise is added from the NOISEX database [16] to test the enhancement process.

We have implemented our proposed design on Xilinx Spartan6 LX45 FPGA using Xilinx system generator platform [18]. The verification of the design was done with clock frequency of 100 MHz (clock period = 10 ns) which is the maximum frequency of our FPGA system with Xilinx Synthesis tool. The time period of the system generator was set to 22.727 ns (=1/44000 s). So, the execution time requirement of the first speech sample through FPGA is 227.27 ns (=22.727 * 10 ns) while the rest of the following samples execute in the pipeline mode one after the other at every clock pulse. For the testing, we adopted 88,001 noisy speech samples. The total execution time taken provided by system generator timing analysis tool process of 88,001 samples is 1.025 s. So, the processing time taken by each of the speech samples through our FPGA is 11.647 ns (=1.025/88,001 s). So, the throughput of our design is 85,910 samples/s (=1/11.647 ns). Due to custom hardware nature of the proposed design, it can work like a coprocessor with other hardware blocks of the system on chip (SoC) architecture. The software execution, i.e., DSP/MCU approach cannot be incorporated with an SoC architecture.

For the evaluation of speech quality perceptual evaluation of speech quality (PESQ) [17] have been applied to the enhanced signals. PESQ scores predicts from −0.5 (*bad*) to 4.5 (*excellent*). In Table 1, we convey the experimental results in terms of PESQ score, estimated time delay and activity of MAD for the different noisy conditions in single and dual microphone scenario. Proposed design provides a good melioration in terms of PESQ score evaluation in each of the noisy conditions. For the validation of the PESQ score of the proposed design, we compare it with our previous works [19, 20] in dual and single microphone scenario respectively. In each condition, PESQ score of the proposed design is similar with [19, 20]. Also, the behavior of microphone activity detector and estimated time delay are correct in each of the microphone conditions. The resource usage of spartan6-Lx45 FPGA of the proposed design is shown in Table 2.

**Table 1** MAD activity and PESQ score

| Noisy source | SNR (dB) | Microphone scenario | MAD | Estimated delay in sec | Proposed PESQ score | PESQ [19] | PESQ[20] |
|---|---|---|---|---|---|---|---|
| White | 0 | Dual | 0 | 0.00289 | 3.23 | 3.22 | |
| White | 3 | Dual | 0 | 0.00168 | 3.21 | 3.21 | |
| White | 0 | Single | 1 | 0 | 3.42 | | 3.42 |
| White | 3 | Single | 1 | 0 | 3.38 | | 3.39 |
| Speech weighted | 3 | Dual | 0 | 0.00288 | 3.30 | 3.30 | |
| Speech weighted | 5 | Dual | 0 | 0.00167 | 3.28 | 3.28 | |
| Speech weighted | 3 | Single | 1 | 0 | 3.39 | | 3.38 |
| Speech weighted | 5 | Single | 1 | 0 | 3.41 | | 3.42 |
| Factory noise | 3 | Dual | 0 | 0.00168 | 3.12 | 3.10 | |
| Factory noise | 5 | Dual | 0 | 0.00288 | 3.08 | 3.06 | |
| Factory noise | 3 | Single | 1 | 0 | 3.22 | | 3.21 |
| Factory noise | 5 | Single | 1 | 0 | 3.18 | | 3.18 |
| Pink | 0 | Dual | 0 | 0.00167 | 3.33 | 3.32 | |
| Pink | 5 | Dual | 0 | 0.00288 | 3.29 | 3.28 | |
| Pink | 0 | Single | 1 | 0 | 3.45 | | 3.44 |
| Pink | 5 | Single | 1 | 0 | 3.49 | | 3.49 |
| Voice babble | 0 | Dual | 0 | 0.00167 | 3.33 | 3.35 | |
| Voice babble | −3 | Dual | 0 | 0.00168 | 3.36 | 3.39 | |
| Voice babble | 0 | Single | 1 | 0 | 3.42 | | 3.48 |
| Voice babble | −3 | Single | 1 | 0 | 3.48 | | 3.55 |
| Multi talker | 0 | Dual | 0 | 0.00167 | 3.22 | 3.20 | |
| Multi talker | 3 | Dual | 0 | 0.00287 | 3.28 | 3.35 | |
| Multi talker | 0 | Single | 1 | 0 | 3.48 | | 3.45 |
| Multi talker | 3 | Single | 1 | 0 | 3.51 | | 3.55 |
| Engine room | 3 | Dual | 0 | 0.00168 | 3.03 | 3.00 | |
| Engine room | 5 | Dual | 0 | 0.00288 | 3.11 | 3.15 | |
| Engine room | 3 | Single | 1 | 0 | 3.17 | | 3.15 |
| Engine room | 5 | Single | 1 | 0 | 3.21 | | 3.26 |

**Table 2** Resource usage of SPARTAN-LX45 FPGA

| Device utilization | Available | Used | Utilization (%) |
|---|---|---|---|
| Slice LUTs | 92,152 | 7998 | 8 |
| DSP48A1S | 180 | 51 | 28 |
| Slice memory | 21,680 | 1099 | 5 |
| Slice registers | 184,304 | 8253 | 4 |
| Bonded IOBs | 296 | 48 | 16 |

## 6 Conclusion

In this paper, we have proposed an FPGA-based speech enhancement system using microphone activity detector which can be used for single as well as the dual microphone scenario. The experimental results show that our proposed design have a good improvement in terms of PESQ score and MAD behavior is correct in each of the microphone conditions. In the future, we wish to do more improvement of PESQ score for the enhanced signals and learning-based speech enhancement machine.

## References

1. Boll, S.: Suppression of acoustic noise in speech using spectral subtraction. IEEE Trans. Acoust. Speech Signal Process. **27**(2), 113–120 (1979)
2. Zhang, Y., Zhao, Y.: Real and imaginary modulation spectral subtraction for speech enhancement. Speech Commun. **55**(4), 509–522. ISSN-0167-6393 (2013)
3. Kamath, S., et. al.: A multi band spectral subtraction method for enhancing speech corrupted by colored noise. In: Proceedings of Acoustics, Speech and Signal Processing, vol. 4 (2002)
4. Adiono, T. et al.: A hardware software co-design for a real-time spectral subtraction based noise cancellation system. In: Proceedings of ISPACS, Nov 2013
5. Biswas, T., et. al.: Audio denoising by spectral subtraction technique implemented on reconfigurable hardware. In: Proceedings of IC3, pp. 236–241 (2014). https://doi.org/10.1109/IC3.2014.6897179.
6. Halupka, D., et. al.: Low power dual microphone speech enhancement using field programmable gate arrays. IEEE Trans. Signal Process. **55**(7), 3526–3535 (2007)
7. Yousefian, N., Loizou, P.C.: A dual-microphone speech enhancement algorithm based on the coherence function. IEEE Trans. Audio Speech Lang. Process. **20**(2), 599–609 (2012)
8. Carter, G.C.: Tutorial overview of coherence and time delay estimation. An applied tutorial for research, development, test, and evaluation engineers, vol. 1, pp. 1–27 (1993)
9. Knapp, C.H., Carter. G.C.: The generalized correlation method for estimation of time delay. IEEE Trans. Acoust. Speech Signal Process. **24**, 320–327 (1976)
10. Champagne, B., et. al.: Performance of time delay estimation in the presence of room reverberation. IEEE Trans. Speech Audio Process. **4**, 148–152 (1996)
11. McAllister, J.: FPGA Based DSP. Springer, US, pp. 363–392. https://doi.org/10.1007/978-1-4419-6345-1_14
12. System Generator DSP User Guide. UG640, 2 Dec 2009. www.xilinx.com/support/sw-manual
13. Matlab/Simulink hardware verifications. www.mathworks.com/products/hdl-verifier
14. Microsoft kinect microphone array. https://msdn.microsoft.com/en-us/library/jj131033.aspx
15. Dual Microphone Database: The 29th Conference on VLSI Design and 15th Conference on Embedded Systems—Design Contest: VLSI 2016 (2016)
16. Varga, A., et. al.: Assessment for automatic speech recognition: II. NOISEX-92: a database and an experiment to study the effect of additive noise on speech recognition system. Speech Commun. **12**(3), 247–251 (1993)
17. Hu, Y., Loizou, P.C.: Perceptual evaluation of speech quality (PESQ), and objective method for end-to-end of speech quality assessment of narrow band telephone network and speech codecs. ITU-T Rec, p. 862 (2000)

18. Hardware Manual Support. www.xilinx.com/support/sw-manual
19. Biswas, T., Mandal, S.B., Saha, D., Chakrabarti, A.: Dual microphone sound source localization using reconfigurable hardware. In: Proceedings of CICBA: Communications in Computer and Information Science, vol. 775. Springer, Singapore (2017)
20. Biswas, T., Mandal, S.B., Saha, D., Chakrabarti, A.: A Novel Reconfigurable Hardware Design for Speech Enhancement Based on Multi-band Spectral Subtraction Involving Magnitude and Phase Components. School Of Information Technology, University of Calcutta (2015)

# Part IV
# Software and Service Engineering

# Optimal Mapping of Applications on Data Centers in Sensor-Cloud Environment

**Biplab Kanti Sen, Sunirmal Khatua and Rajib K. Das**

**Abstract** In sensor-cloud environment, sensing-as-a-service (**Sen-aaS**) is an emerging service paradigm that allows on-demand provisioning of sensor resources as a service in a pay-per-use model. For each application, a disjoint set of virtual sensors (VS) are consolidated in a collaborative wireless sensor network (WSN) platform distributed across the globe. The virtual sensor network (VSN) of an application, formed using VSs, may span across multiple WSNs and the base station for each of these WSNs are placed in a host on a nearest cloud data center (DC). Here, sensor cloud plays the key role to conglomerate the data from various VSs, store them in different hosts, and transmit the same to end user application as a service (Sen-aaS). In this work, we address the problem of mapping applications on the hosts that conglomerate data from various VSs and transmit it to the end user as a constraint optimization problem. The main motivation is to minimize the maximum data migration time of all applications on sensor cloud while satisfying the host's load-balancing constraint. We have proposed an algorithm which can solve the problem optimally under certain conditions. For the general case, if the load-balancing constraint is somewhat relaxed, the maximum delay obtained by our algorithm is optimal. When the load-balancing constraint is to be strictly satisfied, the cost of our solution is slightly more than the optimal provided by an Integer Linear Program.

**Keywords** Sensor cloud · Sensing-as-a-service · Data center

B. K. Sen (✉) · S. Khatua · R. K. Das
Department of Computer Science and Engineering, University of Calcutta, Kolkata, West Bengal, India
e-mail: bksen.cu@gmail.com

S. Khatua
e-mail: sunirmal.khatua.in@ieee.org

R. K. Das
e-mail: rajib.k.das@ieee.org



R. Chaki et al. (eds.), *Advanced Computing and Systems for Security*, Advances in Intelligent Systems and Computing 897,
https://doi.org/10.1007/978-981-13-3250-0_10

## 1 Introduction

Over the last few years, the convergence of WSN and cloud computing technology has emerged as an important computing paradigm called sensor-cloud platform [1]. This platform allows end users to envision the sensor nodes as cloud resources rather than typical sensors. Clearly, sensor cloud provides the advantages of multi-tenancy, pervasiveness, on-demand, self-service and elasticity to a sensor application within a tradition WSN environment.

The physical sensors in sensor-cloud environment are shared among multiple applications through the consolidation of VSs within the physical sensors as shown in Fig. 1 following the sensor-cloud model proposed in [1]. The base station of a WSN provider is placed in a physical host of a cloud data center. This host stores the sensed data from all the VSs placed in the physical sensors of the corresponding WSN. As proposed in Fig. 1, the VSs are grouped together to form a virtual sensor network (VSN) for a particular application. The stored data of the VSs of a VSN are conglomerated in a conglomerator VM before sending it to the end user. These VMs are also placed in a physical host of a cloud data center.

In sensor-cloud platform, an end user requests for Sen-aaS in the form of high-level requirement through sensor modeling language (SensorML) equipped templates. For every Sen-aaS request of a particular application, a set of physical sensors are allocated. Each of the set of allocated physical nodes, serving a particular application, emulates a VS. The set of VSs serving a particular application may span across multiple geographical regions. Clearly, the BSs which collect data from VSs of

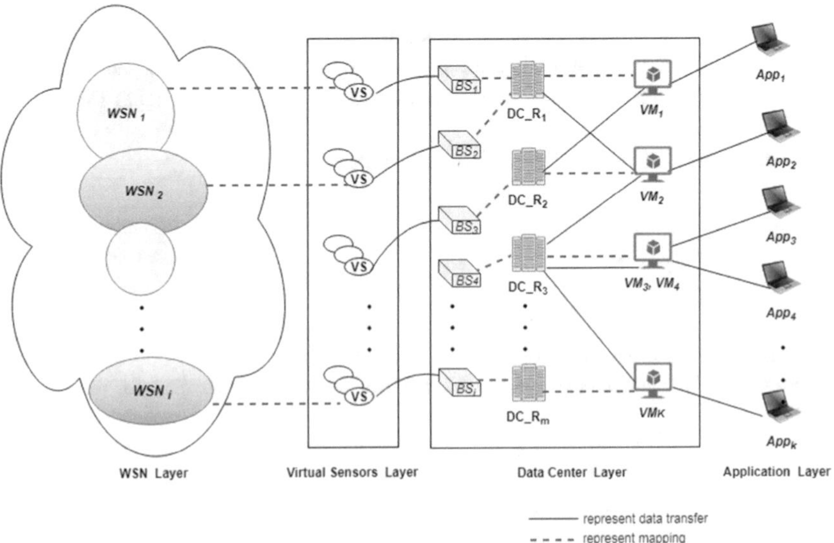


**Fig. 1** Data collection in a sensor-cloud environment using Sen-aaS

a particular Sen-aaS application may also span across geographically distributed DCs. Thus, for each Sen-aaS application, we need to migrate the sensed data from those distributed DCs to a conglomerator VM residing within a particular DC before delivering it to the end user.

Minimizing the data latency in VSN is an important research challenge to be considered in sensor-cloud platform. Clearly, the mapping of the VSs to the WSN and the mapping of the Conglomerator VM to a host in the DC determine the overall data latency of a Sen-aaS application. The mapping of VSs to WSN depends on the physical topology of the WSN as well as the set of target locations of the Sen-aaS application. In this paper, we consider the mapping of conglomerator VM to a DC, and we model the mapping problem as a constraint optimization problem. The main motivation is to minimize the data migration cost that directly affect the service delay of an application while satisfying the DC load-balancing constraint.

## 2 Problem Description

The problem scenario considers one or more end user applications requesting for various types of sensor data from different regions in the form of Sen-aaS. In Sen-aaS, end user applications request for virtual sensors through web templates. The sensor-cloud service provider allocates physical sensors and forms virtual sensors. The sensed data are served to the end users as a service. We assume that there are $n$ hosts $\{h_1, h_2, \ldots, h_n\}$ placed in several DCs. The latency between the hosts is stored in a matrix $L$, i.e., $L_{ij}$ is the latency between $h_i$ and $h_j$. We assume that there are $m$ applications $a_1, a_2, \ldots, a_m$ currently deployed in the sensor cloud. Each application $a_i$ is characterized by the following two parameters:

$f_i = j$ if the final host at which the data for application $a_i$ is to be sent is $h_j$.
$s_i = \{x\text{: } h_x \text{ is a host where the BS for application } a_i \text{ is placed}\}$

Here, $s_i$ ise the set of hosts which act as a source of data for application $a_i$ and $f_i$ is the final destination of the conglomerated data of application $a_i$.

Now, the application mapping problem can be formally stated as follows:

**Definition 1** *Given a set of hosts $H = \{h_1, h_2, \ldots, h_n\}$, the latency matrix $L$, set of applications $A = \{a_1, a_2, \ldots, a_m\}$, and parameters $f_i$, $s_i$ for $i = 1, 2, \ldots, m$, select the host where the data for each application will be conglomerated such that maximum of the time taken by all applications in delivering sensed data to the user is minimized and each host have at most $\lceil m/n \rceil$ and at least $\lfloor m/n \rfloor$ applications.*

We have to determine a mapping $M$: $\{1, 2, \ldots, m\} \rightarrow \{1, 2, \ldots, n\}$ where

$M(i) = j$ implies that application $a_i$ is to be hosted by $h_j$ and the set of applications mapped on host $h_j$ is denoted as $ld_j = \{i\text{: } M(i) = j\}$. The load on host $h_j$ is $|ld_j|$ and we want to make $|ld_j|$ for all $j$ nearly equal to each other. We define the latency for an application as follows:

**Definition 2** *The latency for an application $a_i$ which is mapped in the host $h_j$ is given by*

$$T(i, j) = \max_{k \in s_i}\{L_{i,j}\} + L_{j,f_i}$$

The first term on the RHS is the delay in collecting the data on $h_j$ and the second term is the delay in delivering the collected data to the final destination. The application mapping problem can be formally stated as

**Definition 3** *Find a mapping $M$: $\{1, 2, \ldots, m\} \rightarrow \{1, 2, \ldots, n\}$ s.t, $\max_{1 \le i \le m}\{T(i, M(i))\}$ is minimized and $[m/n] \le |ld_j| \le [m/n]$ for all i.*

## 3 Algorithm for Mapping

In this section, we first present an algorithm where only the condition $ld_j \le [m/n]$ is satisfied but the other condition, i.e., $ld_j \ge [m/n]$ may be violated. The algorithm still attains perfect load balancing if either $n$ divides $m$ or $(m \bmod n) = n - 1$. The algorithm takes as input an $m \times n$ Delay matrix $D$ where $D_{ij} = T(i, j)$, $i = \{1, 2, \ldots, m\}$ and $j = \{1, 2, \ldots, n\}$. $D_{ij}$ is the delay for application $a_i$ if it is hosted in $h_j$. The output of the algorithm is the mapping $M(i)$, $i \in \{1, 2, \ldots, m\}$ and the maximum delay $\max_{1 \le i \le m} \{T(i, M(i))\}$.

```
input : n : Number of applications, m: Number of hosts, Delay matrix D
output: Mapping M(i), maximum delay B
begin
    N ← n × ⌈m/n⌉
    R_i ← min_j D_ij for i = 1, 2, ··· m
    C_j ← min_i D_ij for j = 1, 2, ··· n
    B ← max(max_i R_i, max_j C_j)
    Generate an array SD of size mn which consists of the elements of D in
    ascending order.
    k ← i where SD[i] = B
    Generate an m × N matrix D′ by replicating ⌈m/n⌉ times each column of
    D .
    repeat
        Generate an m × N binary matrix X from D′ where
        X_ij = 1 if D′_ij ≤ B and X_ij = 0 if D′_ij > B
        Find a maximum cardinality matching P on the bipartite graph
        corresponding to X
        if |P| < m then
            k ← k + 1; B ← D[k]
        end
    until |P| = m
    for i = 1 ··· m do
        M(i) ← ⌊(P(i)−1)/⌈m/n⌉⌋ + 1
    end
end
```

**Algorithm 1:** Minimum Delay Mapping

To map application on to hosts, we take the help of bipartite matching which is one-to-one. But we are allowing multiple applications to map to a single host.

That is why we replicate the columns of matrix $D$ $[m/n]$ times as if we have $[m/n]$ copies of each host, and then apply maximum cardinality matching on the bipartite graph represented by matrix $X$. Since each application must be mapped to a host minimum delay for application $a_i$ is $R_i$. Also, if $m \geq n$, each host must get at least one application and the minimum delay for any application mapped to host $h_j$ is $C_j$. Thus, maximum delay for any application is at least $B = \max(\max_i R_i, \max_j C_j)$. The output of matching is a map $P$ where $i$th row of $X$ is mapped to $P(i)$th column of $X$ and $|P|$ is the cardinality of the matching. If $|P| = m$ we are done, otherwise, we try with the next element of $SD$ as $B$. Since the columns $(j-1)[m/n]+1$ to $j[m/n]$ of $D'$ are replication of $j$th column of $D$, all the applications $a_i$ for which $(j-1)[m/n] < P(i) \leq j\,[m/n]$ are mapped to host $h_j$, $j = 1, 2, \ldots, n$.

**Lemma 1.** *When Algorithm 1 terminates with a mapping M, then max$_i$(T(i, M(i)) is equal to the value of B at termination. There does not exist any other mapping M′ such that max$_i$(T(i, M′ (i))) is less than B. Also Algorithm 1 does not map more than $[m/n]$ applications to any single host.*

To keep the algorithm simple, we have employed a linear search on the array $SD$. The time complexity of generating $SD$ is $|SD|\log_2 |SD|$ and that of maximum bipartite matching is $O(m^2 n)$ [2]. Hence, the total time complexity of Algorithm 1 is $O(m^2 n|SD|)$. Instead, a binary search can be employed which will reduce the time complexity to $O(m^2 n \log_2 |SD|)$.

## 3.1 Mapping with Load Balancing

The previous algorithm allows each host to have at most $[m/n]$ applications but if $(m \bmod n) = r$, for perfect load balancing, we can allow only $r$ hosts to have $[m/n]$ applications and the remaining $n - r$ hosts to have $[m/n]$ applications. The following heuristic algorithm is used to determine how the columns of $D$ should be replicated to achieve this.

The following example illustrates how we apply Algorithm 2 to create $D'$ from $D$.

*Example 1:* Consider 10 applications and 3 hosts. The delay matrix $D$ is such that $q_1 = 2$, $q_2 = 5$ and $q_3 = 3$. That is, host $h_1$ gives the least delay for 2 applications, host $h_2$ gives the least delay for 5 applications, and host $h_3$ gives the least delay for 3 applications. So, $q_2 \geq q_3 \geq q_1$. The values of $j_1$, $j_2$, and $j_3$ are thus 2, 3, and 1, respectively. We want to balance the load such that $h_2$ gets $[10/3] = 4$ applications and $h_1$ and $h_3$ get $[10/3] = 3$ applications each. That is achieved by creating matrix $D'$ from $D$ as follows. We replicate column 2 of $D$ 4 times and column 1 and 3 of $D$ 3 times.

We can summarize the use of the two algorithms in this section as follows. If $n$ divides $m$ then, we need to apply only Algorithm 1. Otherwise, we just modify Algorithm 1 by replacing the line **8** by a call to **Algorithm 2** to generate $D'$.

Also, while generating $M(i)$ from $P(i)$, suitable modifications have to be carried out taking into consideration the sequence $(j_1, j_2, ..., j_n)$ and the number of times a particular column is replicated.

**input** : $m$ : Number of applications, $n$: Number of hosts, Delay matrix $D$
**output**: $m \times m$ matrix $D'$
**begin**
  $r \leftarrow m \mod n$
  $g_i \leftarrow \{j : D_{ij}$ is minimum over all $j\}$ for $i = 1, 2, \cdots, m$
  $q_j \leftarrow |\{i : g_i = j\}|$ for $j = 1, 2 \cdots, n$
  Generate a permutation of $1, 2, \cdots, n = j_1, j_2, \ldots, j_n$
  such that $q_{j_1} \geq q_{j_2} \geq \cdots \geq q_{j_n}$
  for $i = 1 \cdots r$ create columns $(i-1)\lceil m/n \rceil + 1 \cdots, i\lceil m/n \rceil$ of $D'$ by replicating coliumn $j_i$ of matrix $D$
  for $i = r+1 \cdots n$ create columns $r\lceil m/n \rceil + (i-r-1)\lfloor m/n \rfloor + 1, \cdots, r\lceil m/n \rceil + (i-r)\lfloor m/n \rfloor$ of $D'$ by replicating coliumn $j_i$ of matrix $D$
**end**

**Algorithm 2:** Load Balance

## 3.2 ILP Formulation

For the ILP, we take as parameter the $m \times n$ Delay matrix $D$ where $D_{ij}$ is the total delay for application $a_i$ if it is run at host $h_j$. The ILP tries to solve binary variables $x_{ij}$, $1 \leq i \leq m$ and $1 \leq j \leq n$ where

$$x_{ij} = \begin{cases} 1, \text{ if } a_i \text{ is mapped to } h_j \\ 0, \, otherwise \end{cases}$$

Now, we can formally state the ILP as

$$\text{minimize } Z$$

subject to the following constraints:

$$\text{(i)} \quad Z \geq D_{i,j} * x_{i,j} \quad \forall_{i,j}$$

$$\text{(ii)} \quad \sum_{j=1}^{n} x_{i,j} = 1 \quad \forall_i$$

$$\text{(iii)} \sum_{i=1}^{m} x_{i,j} \leq [m/n] \; \forall j$$

$$\text{(iv)} \sum_{i=1}^{m} x_{i,j} \geq [m/n] \; \forall j$$

The first constraint ensures that $Z$ is the maximum delay overall applications. The second constraint ensures that an application is mapped to exactly one host. The third and fourth constraint ensure that no host gets more than $[m/n]$ and less than $[m/n]$ applications.

## 4 Experimental Results

We have run a series of experiments to evaluate the performance of the proposed algorithm. In a typical sensor-cloud environment, the number of hosts and applications may be quite large. Even though the ILP gives the optimal solution it is not possible to execute the ILP in a reasonable time when the problem size is large. So, we have two sets of results: one, where the problem size is small and the performance of the proposed algorithm can be compared with that of ILP; and the other, where problem size is large so that executing the ILP is not feasible. Even for the later case, we can find a lower bound on the maximum latency obtainable which can be compared with that given by the proposed algorithm.

### *4.1 Problem Size Is Small*

In this set of experiments, the number of hosts is varied from 30 to 50. Similarly, the number of applications is varied from 100 to 150. The parameters $f_i$ and $s_i$ for each application $a_i$ are chosen randomly. Each element of the latency matrix $L$ takes any random value between 50 and 200 ms. For a given number of hosts, we have plotted the result obtained by ILP, our proposed algorithm, and two greedy heuristics. The greedy heuristics look at the elements $D_{ij}$ for each application $a_i$ and select the host $h_j$ for which $D_{ij}$ is minimum provided that the load-balancing constraint is not violated. In case, mapping an application to that host violates load-balancing constraint, it simply tries the host corresponding to the next minimum. The two variants of the greedy heuristics are termed as Greedy-LF (less fair) and Greedy-MF (more fair). In Greedy-LF, it is only ensured that none of the hosts get more than $[m/n]$ hosts. In Greedy-MF, it is ensured that exactly ($m$ mod $n$) hosts get $[m/n]$ applications and the rest get $[m/n]$ applications. The results of the experiments with 30, 40, and 50 hosts are plotted in Figs. 2, 3 and 4, respectively. From the figures, it is observed that the ILP gives best solution and the greedy heuristic the worst and the proposed algorithm solutions are quite close to those of ILP. Also, Greedy-LF gives lower latency at the cost of poorer load distribution. Since for each different set up (changing number of applications or hosts) the parameters $f_i$, $s_i$ and latency matrix $L$ are generated afresh randomly, the total delay sometimes decreases with increase in number of applications. Since the solutions of ILP also show similar pattern, this is not due to inconsistency of the algorithm but due to randomness of the parameters. When the number of applications is an integral multiple of the number of hosts, we need

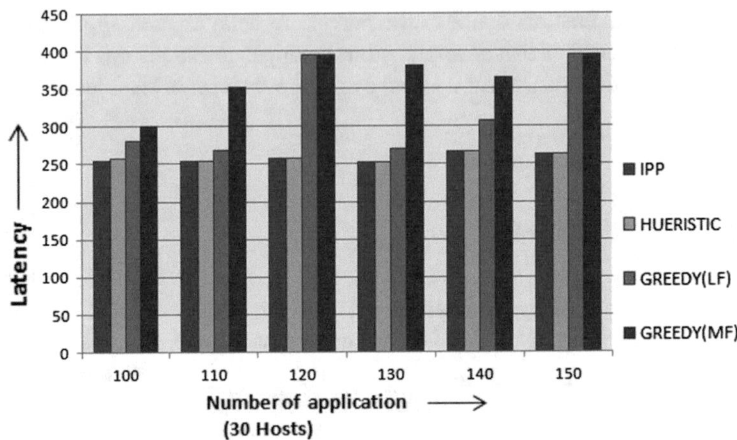


**Fig. 2** Maximum delay for 30 hosts

to use only Algorithm 1 and the performance of the proposed algorithm is optimal (same as given by ILP). We can see that for the following # of applications, # of hosts pair, i.e., (120, 30), (150, 30), (120, 40), (100, 50), and (150, 50), the maximum delay for ILP and proposed algorithm are identical (Greedy-LF and Greedy-MF give equal latency in such cases). Only in the cases where the number of application is not an integral multiple of the number of hosts, we need to use Algorithm 2 for load balancing before using the maximum bipartite matching. It is to be noted that Algorithm 1 obeys the load-balancing constraint to some extent in the sense that no host gets more than $[m/n]$ applications. But, if $(m \bmod n) = r$, simply by using Algorithm 1 it may happen that more than $r$ hosts have $[m/n]$ applications. In that case the remaining hosts may end up with less than $[m/n]$ applications which is undesirable. To avoid such a result, we have designed the heuristic Algorithm 2 and then the maximum delay is not guaranteed to be the same as given by the ILP. But the experimental results show that in such cases the cost of our proposed algorithm is either same as that given by ILP or just a little bit more. Thus from the results, we can conclude that Algorithms 1 and 2 can provide proper load balancing and also keep the maximum delay overall applications quite close to the optimal.

## *4.2 Problem Size Is Large*

In this set of experiments, the number of applications is varied from 160 to 640 and number of hosts is fixed at 100. We have considered the performance of three algorithms: (i) the proposed Algorithm **A:1–2** (Algorithm 1 with Algorithm 2 to generate $D'$) (ii) only Algorithm 1 **A:1** and (iii) Greedy-MF. Algorithm **A:1** alone does not ensure proper load balance but at the same time it gives a lower bound on

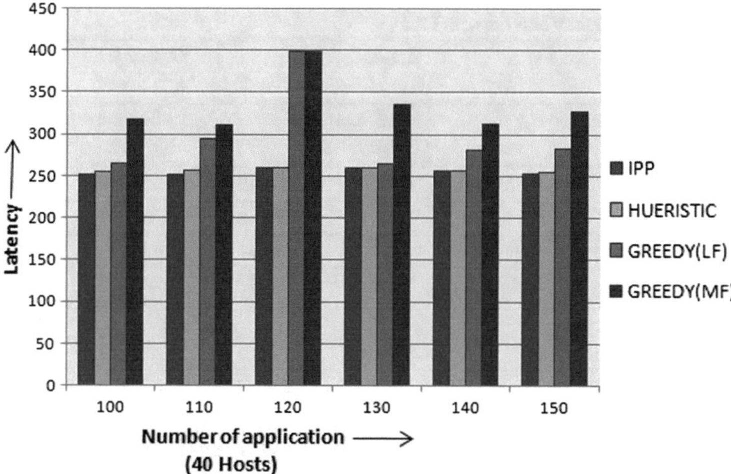


**Fig. 3** Maximum delay for 40 hosts

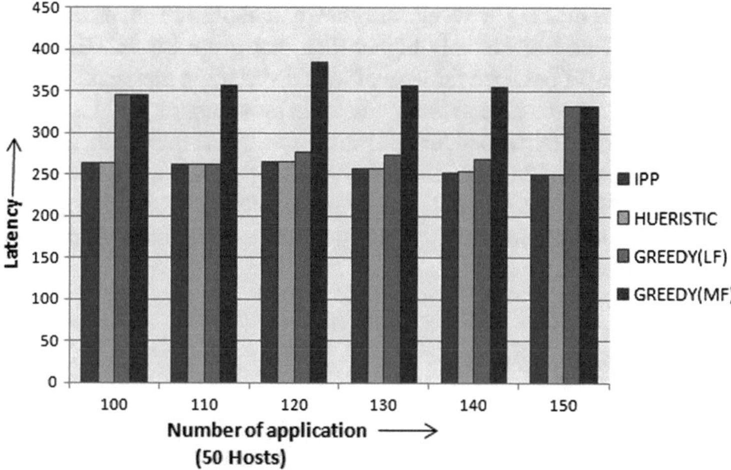


**Fig. 4** Maximum delay for 50 hosts

the maximum latency according to Lemma 1. Thus, even though we are not using ILP we can be assured that the maximum latency that ILP would give cannot be less than that given by Algorithm **A:1**. To measure how fair the load balance is, we use Jain's fairness index [3] which is defined for $n$ numbers $x_1, x_2, \ldots, x_n$ as

$$F = \frac{\left(\sum_{i=1}^{n} x_i\right)^2}{n\sum_{i=1}^{n} x_i^2}$$

**Table 1** Maximum latency and fairness index

| # of applications | A:1 | | A:1–2 | | Greedy | |
|---|---|---|---|---|---|---|
| | Maximum latency | *F* | Maximum latency | *F* | Maximum latency | *F* |
| 160 | 253.8 | 0.83338 | 254.2 | 0.9143 | 331 | 0.9143 |
| 320 | 252.2 | 0.84858 | 255.8 | 0.9846 | 336.6 | 0.9846 |
| 480 | 252.6 | 0.97748 | 254.8 | 0.9931 | 365 | 0.9931 |
| 640 | 253.2 | 0.9566 | 253.4 | 0.9942 | 354.4 | 0.9942 |

For application mapping problem, the value of $x_i$, $i = 1, 2, \ldots, n$ is the number of applications mapped to host $h_i$ or the load of host $h_i$. Maximum possible value of $F$ is 1 when all $x_i$'s are equal. A small value of $F$ indicates that the load is unevenly distributed.

Here, for a given number of applications, we have run 10 experiments with different seeds for random number generators and computed the average value of maximum latency and fairness index. The results are shown in Table 1. The maximum latency for the Greedy-MF is considerably higher than that given by **A:1–2**. We can see that there is a trade-off between fairness of load distribution and maximum latency. As the number of applications $m$ is not divisible by number of hosts $n$, we cannot expect a value of 1 for fairness index. **A:1** gives the lowest maximum latency at the cost of lower fairness index. Also, the fact that **A:1** gives a lower bound on optimal latency and **A:1–2** have maximum latency only slightly more than that given by **A:1** demonstrates that the proposed algorithm gives close to optimal latency with best possible fairness index.

## 5 Related Works

A number of research works have been conducted toward the detailed architecture and ideology of sensor cloud over the last few years [1, 4–7]. However, very few of the research works have explored the networking aspects and the challenges behind the sensor-cloud infrastructure [1, 4, 8]. On the other hand, the majority of the earlier works lack the realization of virtualization [9–11] which is the key technology that maps virtual sensors with the physical sensors. Works in [9–11] suggested the integration of wireless sensor technology to cloud computing. The authors in [12] inscribed the problem of dynamic gateway allocation while transmitting the sensed data from the networks to the cloud. Virtualized platform SenShare presented in [4], addressed the technical challenges in transforming sensor networks into open access infrastructures capable of supporting multiple co-running applications. Some detailed aspects of virtualization and the sensor data management also presented in

[6, 7]. Work in [5, 13] identified the principal benefits, and challenges involved in implementation of sensor-cloud infrastructure.

Subsequently, few research works have explored some of the technical aspect and different usage models of sensor-cloud platform. In [1] a collaborative model has been presented to implement sensor-cloud platform and the problem of covering set of locations of interest (targets) by this collaborative platform has also been explored. Another work [8] addresses the problem of scheduling a particular DC that congregates data from various VSs in sensor-cloud platform. The scheduler process is designed under constraints to ensure user satisfaction and maintenance of QoS, simultaneously. Nguyen and Huh discussed the security aspects of sensor cloud in [14]. In [15], a scenario of multiple target tracking was explored to examine the implementation of sensor cloud from an application perspective. The problem of consolidations of VMs under QoS constraints addressed in [16, 17]. This work proposes an efficient virtualization scheme for physical sensor nodes and seeks for an optimal location-based composition of virtual sensors.

Different pricing models and usage models for the sensor cloud are proposed in [18].

## 6 Conclusion

The proposed work is focused on finding the optimal mapping of applications to hosts in a sensor-cloud infrastructure. In a sensor cloud, data from various physical sensors are grouped to form a VS. An application's data may come from multiple such VSs spanning several WSNs. Therefore, given a set of VSs and the final destination host for each application, making an optimal selection of host to run each application, is of practical interest. We have proposed an algorithm using matching on bipartite graphs which provided minimum delay while ensuring that no host gets a load of more than that which is proper ($[m/n]$ for $m$ applications and $n$ hosts). We have then extended that algorithm to ensure that no host get less than $[m/n]$ applications at the cost of getting delay slightly more than the optimal. The trade-off between fairness of load distribution and maximum delay is also substantiated with experimental results. For a small-sized problem, ILP can be used to get a mapping which ensures minimum delay and proper load balancing. For large values of $m$ and $n$, ILP becomes impractical and in such situations, the algorithm proposed in this paper can be relied on to get a close to optimal mapping.

## References

1. Sen, B.K., Khatua, S., Das, R.K.: Target coverage using a collaborative platform for sensor cloud. In: 2015 IEEE International Conference on Advanced Networks and Telecommunications Systems (ANTS), pp. 1–6. IEEE (2015)

2. Cormen, T.H., Leiserson, C.E., Rivest, R.L., Stein, C.: Introduction to Algorithms, 3rd edn. The MIT Press, Cambridge, Massachusetts (2009)
3. Jain, R., Chiu, D.M., Hawe, W.: A quantitative measure of fairness and discrimination for resource allocation in shared computer systems (PDF). DEC Research Report TR-301 (1984)
4. Leontiadis, I., Efstratiou, C., Mascolo, C., Crowcroft, J.: SenShare: transforming sensor networks into multi-application sensing infrastructures. In: European Conference on Wireless Sensor Networks, pp. 65–81. Springer, Berlin, Heidelberg (2012)
5. Alamri, A., Ansari, W.S., Hassan, M.M., Hossain, M.S., Alelaiwi, A., Hossain, M.A.: A survey on sensor-cloud: architecture, applications, and approaches. Int. J. Distrib. Sens. Netw. **9**(2), 917923 (2013)
6. Yuriyama, M., Kushida, T., Itakura, M.: A new model of accelerating service innovation with sensor-cloud infrastructure. In: 2011 Annual SRII Global Conference (SRII), pp. 308–314. IEEE (2011)
7. Madria, S., Kumar, V., Dalvi, R.: Sensor cloud: a cloud of virtual sensors. IEEE Softw. **31**(2), 70–77 (2014)
8. Chatterjee, S., Misra, S., Khan, S.: Optimal data center scheduling for quality of service management in sensor-cloud. IEEE Trans. Cloud Comput. (2015)
9. Tan, K.L.: What's next?: sensor+cloud!? In: Proceedings of the Seventh International Workshop on Data Management for Sensor Networks, pp. 1–1. ACM (2010)
10. Zhang, P., Yan, Z., Sun, H.: A novel architecture based on cloud computing for wireless sensor network. In: Proceedings of the 2nd International Conference on Computer Science and Electronics Engineering. Atlantis Press (2013)
11. Hassan, M.M., Song, B., Huh, E.N.: A framework of sensor cloud integration opportunities and challenges. In: Proceedings of the 3rd International Conference on Ubiquitous Information Management and Communication, pp. 618–626. ACM (2009)
12. Misra, S., Bera, S., Mondal, A., Tirkey, R., Chao, H.C., Chattopadhyay, S.: Optimal gateway selection in sensor cloud framework for health monitoring. IET Wirel. Sens. Syst. **4**(2), 61–68 (2013)
13. Sheng, X., Tang, J., Xiao, X., Xue, G.: Sensing as a service: challenges, solutions and future directions. IEEE Sens. J. **13**(10), 3733–3741 (2013)
14. Nguyen, T.D., Huh, E.N.: An efficient key management for secure multicast in sensor-cloud. In: 2011 First ACIS/JNU International Conference on Computers, Networks, Systems and Industrial Engineering (CNSI), pp. 3–9. IEEE (2011)
15. Chatterjee, S., Misra, S.: Target tracking using sensor-cloud: sensor target mapping in presence of overlapping coverage. IEEE Commun. Lett. **18**(8), 1435–1438 (2014)
16. Beloglazov, A., Buyya, R.: Managing overloaded hosts for dynamic consolidation of virtual machines in cloud data centers under quality of service constraints. IEEE Trans. Parallel Distrib. Syst. **24**(7), 1366–1379 (2013)
17. Chatterjee, S., Misra, S.: Optimal composition of a virtual sensor for efficient virtualization within sensor-cloud. In: 2015 IEEE International Conference on Communications (ICC), pp. 448–453. IEEE (2015)
18. Chen, S.L., Chen, Y.Y., Hsu, C.: A new approach to integrate internet of-things and software-as-a-service model for logistic systems: a case study. Sensors **14**(4), 6144–6164 (2014)

# Constraint Specification for Service-Oriented Architecture

**Shreya Banerjee, Shruti Bajpai and Anirban Sarkar**

**Abstract** Constraint specification is an essential and significant part in Service-Oriented Architecture (SOA). Expressive specifications of constraints are highly required for effective accomplishment of the distinct phenomenon in SOA like service discovery, execution of service and service composition. However, the precise specification is only obtained through well-formed descriptions of both syntax and semantics. Yet, the majority of existing approaches has described constraints focusing on well-defined syntax. Thus, representations of rigorous semantics are absent in major numbers of the existing constraint specifications related to SOA. To address this issue, in this paper, an ontology-based constraint specification is proposed for SOA that can be applied in three aspects—service discovery, service execution and service composition. These proposed constraints are specified in meta-model level. Further, the proposed approach is implemented using Protégé. Moreover, it is also illustrated using a suitable case study to prove its practical usability.

**Keywords** Service-oriented architecture · Meta-model · Constraints
Service composition · Service discovery · Service execution · SOA

## 1 Introduction

Service-Oriented Architecture (SOA) is a paradigm that is used for designing, developing, deploying and maintaining the systems that involve services as the key element. An SOA reference model defines the essence of SOA and emerges with a

S. Banerjee (✉) · S. Bajpai · A. Sarkar
Department of Computer Science and Engineering, National Institute of Technology, Durgapur, India
e-mail: shreya.banerjee85@gmail.com

S. Bajpai
e-mail: bajpai.shruti1705@gmail.com

A. Sarkar
e-mail: sarkar.anirban@gmail.com



R. Chaki et al. (eds.), *Advanced Computing and Systems for Security*,
Advances in Intelligent Systems and Computing 897,
https://doi.org/10.1007/978-981-13-3250-0_11

vocabulary and a common understanding of SOA [1]. A service delivers business functionalities. It can be reused over several systems. Distinct services compose to form a system either in orchestration or in choreography through interfaces.

SOA is a very complex architectural style [2]. Thus, many challenges exist to model this complex architecture effectively. To overcome these challenges, ontology is used in the existing approaches. Since, mixing of ontology with conceptual modelling is a better solution for getting a precise conceptual model [3]. In [4], such an ontology-driven SOA meta-model is described. This conceptualization [4] addresses challenges—(1) how to cover both behavioural, structural characteristics and their interdependency of SOA and (2) how to support both businesses understandable and executable features of SOA. Yet, several challenges still exist that must be addressed in order to make this meta-model [4] effective. Accurate specifications of constraints related to SOA are crucial among them.

Constraints related to SOA can be defined as restrictions whose meaningful interpretation is necessary in SOA-related facets such as services, services operations, provider, requester, service interface, service description, etc. [5]. Hence, service operations, binding between the service provider and requester, service invoke, composition between services—these distinct aspects of SOA can be changed based on specifications of related constraints. Thus, constraint specifications need to be accurate and precise so that distinct aspects of SOA can be accomplished effectively. Consequently, well-formed description of both syntax and semantics in constraint specification is the prime requisite. Yet, the majority of existing approaches focus on well-defined syntax. Lack of precise semantics in constraint specification create serious drawbacks those may affect distinct aspects of SOA.

Ontology can help to make the specification of constraints well-formed and meaningful. In [6], ontology is defined as an explicit specification of shared conceptualization in terms of concepts, relationships present between those concepts and related axioms. Axioms may be expressed formally to provide precise semantics. Thus, ontology can play a vital role in the validation of specification also. Further, ontology can be specified in level hierarchies—upper level, middle level and lower level. Upper level ontology provides most generic semantics independent of any domain. Middle-level ontology expresses domain specific semantics and conforms towards an upper level ontology. Further, lower level ontology provides application-specific semantics for a particular domain and conform towards middle-level ontology of that domain [3, 7]. In this paper, the proposed ontology is middle-level ontology which is specifically used to define the meta-model of software service domain. It provides a vocabulary for constructs of software service domain. Further, from this vocabulary, meaningful semantics for constructs of different applications in software service domain can be derived.

This paper is aimed for addressing the afore-mentioned necessity of concise and precise syntax and semantics in constraint specification related to SOA. For this purpose, this paper extends the meta-model described in [4] with ontology-based specification of constraints. The novelty of the proposed work is that the proposed constraints are related to three important aspects of SOA—service discovery, service execution and service composition. In the proposed meta-model, constraints are normally specified with the triggering events and encapsulated within the structural part of SOA. Specified constraints can be generated in two ways: (i) statically pre-imposed at the time of service definition (user specified) and (ii) raised dynamically by the system itself (system generated). System-generated constraints can be used by service broker to register, publish and discover services. A supporting ontology incorporates all the constraint specifications. Since the proposed specification is based on meta-model level, it can also be applied towards large numbers of domain based on SOA. Further, the whole specification is in mathematical logic. Beside this, the proposed specification is implemented in an ontology tool Protégé [8] for initial validation of proposed conceptualization. Moreover, a case study is used to prove the practical usability of the proposed specification.

The remainder of this paper is organized as follows. Section 2 throws light on the related work. Section 3 includes a brief description of ontology-driven SOA meta-model described in [4]. In Sect. 4, ontology-based constraints are proposed and implemented using Protégé. Further, in Sect. 5, the proposed specification is illustrated using a suitable case study. Finally, Sect. 6 concludes this paper with future scope.

## 2 Related Work

A lot of research work exists in the field of constraint specification related to SOA. In [2], the authors have described a pattern-based constraint description approach for web services. In this research work, a constraint pattern hierarchy, which focuses on the messages exchanged between client and server, is discussed. Besides this, an RDF-based constraint description language is described to give explicit semantics to the specified constraint descriptions. In [9], an extension is specified in the functionality of the web services broker to include constraint specification and processing, which enables the broker to find a good match between a service provider's capabilities and a service requestor's requirements. These extensions are made to the Web Services Description Language to include constraint specifications in service descriptions and requests. In [10], the authors have presented a consistency-based service composition approach, where composition problems are modelled in a generative constraint-based formalism. This approach works for the dynamic service composition problem. In [11], conceptual graphs are used to represent constraint specification problems. Further, various constraints over knowledge-intensive domains are presented with formalism. In [12], the authors have described a framework that consists of Process Constraint Ontology and Process Constraint Language to handle constraints. It addresses the problems of constraint specification in workflow processes using an extensible ontology. This work connects the constraints with various process elements like activities, performers and data. In [7], a foundation for service computing and service science is defined with a core reference ontology of service known as UFO-S (Unified Foundational Ontology). The authors have focused on three phases of service life cycle—service offer, service negotiation and service delivery. Constraints are represented through logical axioms. However, principal concepts related to SOA are represented based on Unified Modelling Language (UML) which is a semiformal language and less efficient than ontology for providing rigorous and consistent descriptions for concepts. In [5], the authors have described a semantic service discovery system based on OWL-S for discovering complex constraint-based services dynamically. They have proposed service constraints using Semantic Web Rule Language (SWRL) and used a rule engine for matching and handling dynamism. However, this approach is based on semantic web services.

Major approaches have concentrated on well-defined syntax but not on semantics. Further, less approaches has focused on meta-model level constraint specification for SOA. Thus, most of the approaches are domain specific and less reusable in vast numbers of the domain.

## 3 Brief Description of Ontology-Driven Meta-model for Service-Oriented Architecture [4]

The ontology-driven meta-model for SOA [4] provides unifying semantics and syntax towards distinct concepts of SOA. The conceptualization is specified in mathematical logic based on axioms of Generalized Ontology Modelling (GOM) described in [3]. GOM is an upper level ontology. Different middle-level ontology related to distinct domains such as software service, health care etc., could be derived from GOM. Further, middle-level ontology can be instantiated towards lower level ontology related to different applications software in distinct domain. Several important concepts of SOA meta-model are specified in Table 1. The three key conceptualizations are *Information Model* (*IM*), *Behavioural Model* (*BM*) and *Action Model* (*AM*). Distinct concepts of SOA and their in-between dependencies or relationships are included to these three models.

*Information Model* (*IM*): It is defined as a conceptualization of structural concepts included in SOA such as *Service* (*S*), *Data* (*DA*), *Service Interface* (*SI*) and *Actor* (*AC*),
*Service Description* (*SD*), *Service Contract* (*SC*), *Registry* (*REG*) and *System* (*SYS*). These concepts are structural since these have certain properties. Thus, *Information Model* (*IM*) can be instantiated from *Structural Element Conceptualization* ($C_{SE}$) of GOM.
*Action Model* (*AM*): It is defined as a conceptualization of functionality and related relationships included in SOA such as operations offered by services. Functionalities offered by services are activities since those functions are related to distinct tasks. Thus, *Action Model* (*AM*) can be instantiated from *Activity Element Conceptualization* ($C_{AE}$) of GOM. *Activity Elements* (*AE*) included in this model is *Actions* (*ACN*).
*Behavioural Model* (*BM*): It is defined as a conceptualization of effects and responding by services towards the surrounding environment. Real-world effects are events

**Table 1** Distinct concepts and conceptualization of SOA meta-model

| Conceptualization | Concepts | Description |
|---|---|---|
| Information Model (*IM*) | Service (*S*) | A package of closely related business functionalities, which are called repeatedly in a similar fashion |
| | Data (*DA*) | Input and output parameters of operations, which constitute a service |
| | Service Interface (*SI*) | Service Interface defines the way in which other elements may interact and exchange information with a service |
| | System (*SYS*) | A system is an organized collection of services. It can be of two types—Choreography System ($SYS_{CHG}$) and Orchestration System ($SYS_{ORC}$) |
| | Choreography System ($SYS_{CHG}$) | This type of System is organized through the means of choreography among distinct services |
| | Orchestration System ($SYS_{ORC}$) | This type of System is organized through the means of orchestration among distinct services |
| Action Model (*AM*) | Action (*ACN*) | Represent activities or operations through which functionality of a service is accomplished |
| | Composite Action (*CACN*) | This is a composition of simple actions |
| | Simple Action (*SACN*) | Atomic activity which has several tasks |
| Behavioural Model (*BM*) | Effect (*Eff*) | Represent the outcome after invoking a service |
| Artefacts (*AFE*) | Properties (*P*) | Properties have denoted the attributes of concepts included in *Information Model* (*IM*) |
| | Messages (*MSG*) | Messages hves denoted the send and receive messages through Service Interfaces |
| | Constraint (*CO*) | Constraint has denoted conditions related to effects |
| | Time (*TM*) | This has specified the existing time of all concepts of the proposed meta-model |
| | Visibility Flag (*Flg*) | Flags have denoted visibility of Services |

since achieving effects depend on distinct constraints. Thus, *Behavioural Model* (*BM*) can be instantiated from *Event Element Conceptualization* ($C_{EE}$) of GOM. Event *Elements* (*EE*) included in this model can be *Effect* (*Eff*).
*Behavioural Model* (*BM*): It is defined as a conceptualization of effects and is responded by services towards the surrounding environment. Real-world effects are events since achieving effects depend on distinct constraints. Thus, *Behavioural Model* (*BM*) can be instantiated from *Event Element Conceptualization* ($C_{EE}$) of GOM. Event *Elements* (*EE*) included in this model can be *Effect* (*Eff*).

In conceptualization of SOA, different concepts are related with each other using distinct relationships. Those relationships are instantiated from relationships of GOM [3]. Several crucial relationships of SOA meta-model are summarized in Table 2.

## 4 Proposed Ontology-Based Constraint Specification

Distinct constraints can be imposed on facets of SOA in different ways and aspects. Service compositions put together loosely coupled component services to perform more complex, higher level or cross-organizational tasks in a platform-independent manner.

In this context, constraints are the restrictions those may impose on distinct facets of SOA when services are composed. Thus, the suitable semantic specification of distinct constraint is a prime requisite so that services can be unambiguously discovered, executed and composed. To address this issue, in this section, constraints are formally specified at service discovery, service execution and service composition level as well as on data related to a service based on ontology theory. In the proposed meta-model, constraints are specified along with the specification of *Event Elements* (*EE*) through *Has Containment* (*HCO*) relationship. Further, those *Event Elements* are embedded within *Structural Elements* (*SE*) through *Inter Containment* ($CTD_{SE\text{-}EE}$) relationship. To realize the source/type of the constraints (user specified or system generated), a Boolean *Flag* can be used with *Constraints* (*CO*).

### *4.1 Proposed Constraints and Related Types*

Distinct constraints are divided into four types—(a) *Ordering constraints* are imposed during service composition, (b) *Visibility constraints* are imposed during service discovery, (c) *Data constraints* are imposed on data and at the time of service execution and (d) *Time constraints* may be imposed in all these aspects. All of these Constraints are specified along with the specification of *Event Elements* (*EE*) through *Has Containment* (*HCO*) relationship described in [4]. *Data Constraints*, *Ordering Constraints* and *Visibility Constraints* are user specified and related flag is set to *0*.

**Table 2** Summarization of distinct relationships of proposed SOA meta-model

| Relationships in proposed conceptualization | Description | Equivalent relationships In GOM |
|---|---|---|
| Has Interface | Between concepts of Service and Service Interface | Association (*AS*) |
| Has Consumed | Between concepts of Service Consumer and Service | |
| Has Provided | Between concept of Service Provider and Service | |
| Has Bind | Between two Service Interfaces | Intra Interaction (*IRS*) |
| Has Find | Between Service Interface and Registry, when the consumer has to find the service | |
| Has Publish | Between Service Interface and Registry, when the provider has published the service | |
| Has AND relationship | Between two actions executed in parallel | |
| Has OR relationship | Between two actions executed by alternatives | |
| Has Sequence | Between two actions executed in sequence | |
| Has Data | Between Service Interface and Data | Intra Containment (*CTS*) |
| Has Composite Action | Between two Composite Actions | |
| Has Simple Action | Between Composite Action and Simple Action | |
| Has Effect | Between Service and its real-world effect | |
| Collaboration | Connection between two Services or two Actors when each has played distinct Roles | Intra Collaboration (*CR*) |
| Send Message | Connection between Actions and Messages | Send Message (*SM*) |
| Receive Message | Connection between Actions and Messages | Receive Message (*RM*) |
| Has Time | Between all concepts and corresponding time | Has Time (*HTM*) |
| Inverse Relationships | Responsible for the addition of related concepts towards Service dynamically | Inverse Relationships (*IRL*) |

On the other hand, *Time Constraint* is generated in the system and related flag is set to *1*.

(a) ***Ordering Constraints***: *Ordering Constraints* are applied at the service composition level. These constraints specify both ordered and unordered arrangement of services in an orchestrator or choreography system. Ordered constraints specify the order or enumeration in which two or more services will compose together to form either an orchestrator system or a choreography system. Based on any ordering constraints, different services compose with each other or with a specific service in a sequence of increasing timestamps. On the other hand, unordered constraints do not specify any order or enumeration in which services will compose. In this case, the composition can be accomplished either in parallel or randomly. Both ordered and unordered constraints can connect distinct services using several logical operators such as AND, OR, X-OR, etc. Further, according to ordered constraints, distinct relationships can exist between services. Formal specifications of those relationships are specified next. However, these relationships cannot exist in case of unordered constraints.

Let, there are four services having interfaces *I1*, *I2*, *I3* and *I4*. An event *EV1* is triggering on *I2*, *I3* and *I4*. Let, *EV1* is specified with distinct ordered constraints and is applied on interfaces *I2*, *I3* and *I4* in timestamps *t2*, *t3* and *t4*, respectively, in order to compose with *I1*. According to ordered constraints, distinct relationships exist between *I2*, *I3* and *I4* are:

(i) *is_next_to*—This relationship defines that *I3* will be composed with *I1* after *I2* is composed with *I1*. In this context, *EV1* is imposed on *I2* and *I3* in time *t2*. It is formally specified as

$$\mathrm{F1}: \forall S1 \exists S2 \exists S3 \exists I1 \exists I2 \exists I3 \exists EV1 \exists EV2 \exists EV3 \exists t1 \exists t2 \exists t3((AS(S1, I1) \wedge$$
$$AS(S2, I2) \wedge AS(S3, I3) \wedge EE(EV1) \wedge CTD_{SE-EE}(I2, EV1) \wedge CTD_{SE-EE}(I3, EV1) \wedge$$
$$HCO\big(EV1, ordered\_constraint_{Flag(0)}\big) \wedge HTM(EV1, t2) \wedge$$
$$equal(t3, Increment(t2, 1)) \rightarrow is\_next\_to(I3, I2) \wedge composed\_with(I1, I2, t2) \wedge$$
$$composed\_with(I1, I3, t3) \wedge logical\_operators(I2, I3)))$$

(ii) *is_last_to*—This relation defines that *I2* will be composed with *I1* before *I3* is composed with *I1*. In this context, *EV1* is imposed on *I2* and *I3* in time *t2*. It is formally specified as

$$F2: \forall S1 \exists S2 \exists S3 \exists I1 \exists I2 \exists I3 \exists EV1 \exists EV2 \exists EV3 \exists t1 \exists t2 \exists t3((AS(S1, I1) \wedge$$
$$AS(S2, I2) \wedge AS(S3, I3) \wedge EE(EV1) \wedge CTD_{SE-EE}(I2, EV1) \wedge CTD_{SE-EE}(I3, EV1) \wedge$$
$$HCO\big(EV1, ordered\_constraint_{Flag(0)}\big) \wedge HTM(EV1, t2) \wedge$$
$$(equal(t2, Decrement(t3, 1)) \rightarrow is\_last\_to(I2, I3) \wedge composed\_with(I1, I2, t2) \wedge$$
$$composed\_with(I1, I3, t3) \wedge logical\_operators(I2, I3)))$$

(iii) *precedes*—This relation defines that *I2* will be composed with *I1* before *I4* is composed with *I1*, at any time not necessarily the immediate previous. In this context, *EV1* is imposed on *I2* and *I3* in time *t2*. It is formally specified as

$$
\begin{aligned}
&F3 : \forall S1 \exists S2 \exists S4 \exists I1 \exists I2 \exists I3 \exists EV1 \exists EV2 \exists EV3 \exists t1 \exists t2 \exists t3((AS(S1, I1) \wedge \\
&AS(S2, I2) \wedge AS(S4, I4) \wedge EE(EV1) \wedge CTD_{SE-EE}(I2, EV1) \wedge CTD_{SE-EE}(I4, EV1) \wedge \\
&HCO\big(EV1, ordered\_constraint_{Flag(0)}\big) \wedge HTM\,(EV1, t2) \wedge lessThan(t2, t3) \wedge \\
&lessThan(t3, t4)) \rightarrow (precedes(I2, I4) \wedge composed\_with(I1, I2, t2) \wedge \\
&composed\_with(I1, I4, t4) \wedge logical\_operators(I2, I4)))
\end{aligned}
$$

(iv) *follows*—This relation defines that *I4* will be composed with *I1* after *I2* is composed with *I1*, at any time not necessarily the immediate next. In this context, *EV1* is imposed on *I2* and *I4* in time *t2*. It is formally specified as

$$
\begin{aligned}
&F4 : \forall S1 \exists S2 \exists S4 \exists I1 \exists I2 \exists I3 \exists EV1 \exists EV2 \exists EV3 \exists t1 \exists t2 \exists t3((AS(S1, I1) \wedge \\
&AS(S2, I2) \wedge AS(S4, I4) \wedge EE(EV1) \wedge CTD_{SE-EE}(I2, EV1) \wedge CTD_{SE-EE}(I4, EV1) \wedge \\
&HCO\big(EV1, ordered\_constraint_{Flag(0)}\big) \wedge HTM\,(EV1, t2) \wedge greaterThan(t3, t2) \wedge \\
&greaterThan(t4, t3)) \rightarrow (follows(I4, I2) \wedge composed\_with(I1, I2, t2) \wedge \\
&composed\_with(I1, I4, t4) \wedge logical\_operators(I2, I4)))
\end{aligned}
$$

(b) ***Data Constraints***—*Data Constraints* (*DCO*) are specifically meant for the data. These are the conditions or rules applied on the data itself. Either this data can be input data or output data related to a service. Data constraints are necessary to put restrictions on the type, length, value etc., of data. There can be four types of data constraints. These are *Boundary Constraints* (*BDCO*), *Length Constraints* (*LDCO*), *Single Value Constraints* (*SDCO*) and *Type Constraints* (*TDCO*).

*Boundary Constraints* define the boundaries within which a data can have value. Therefore, there is a minimum boundary below which the data cannot have the value and a maximum boundary above which the data cannot have the value. *Length Constraints* define the length of data violating which causes error. These length constraints can be put either to a minimal or to a maximum. Thus, the data cannot have a length less than *x* characters or a length greater than *y* characters. Here, *x* and *y* are simple variables set by the designer. *Single Value Constraints* are meant for the data when the data holds a single value at a time rather than a set of values. This type of constraint is important when evaluating the conditions or giving the status or entering of some choice by the user. Lastly, *Type Constraints* depict what should be the type of data. It can be any of inbuilt or user-defined data types. This constraint is important, as it is a major factor during operations, which are delivered by a particular service and execute on related data. Further, it is also essential that the input and output types of data should be compatible and if not so, they are made compatible using type casting. Thus, data constraints can restrict a particular service execution and distinct data too related to a service. The formalization of data constraints is specified as

$$F7 : \forall S1 \exists I1 \exists d \exists dc \exists EV1 (S(S1) \wedge SI(I1) \wedge DA(d) \wedge DCO(dc) \wedge AS(S1, I1) \wedge CTS_{SE-SE}(I1, d) \wedge EE(EV1) \wedge CTD_{SE-EE}(d, EV1) \wedge HCO\big(EV1, dc_{Flag(0)}\big) \rightarrow (BDCO(dc) \vee TDCO(dc) \vee SDCO(dc) \vee LDCO(dc)))$$

(c) ***Time Constraints***—Time Constraints (*TCO*) are related to timing aspects of the events. They are imposed on the events when some temporal aspect is considered. Three types of *Time Constraints* are proposed. These are *Frequency* (*FTCO*), *Interval* (*ITCO*) and *Response* (*RTCO*) *Time Constraints*. *Frequency Time Constraints* are meant to impose restrictions on the number of times of occurrence of an event. Those tell how frequently an event has been triggered or some message has been passed. *Interval Time Constraints* restrict the existing time duration of an event. *Response Time Constraints* define the time that a service will take to response an event. The formalization of time constraints is

$$F8 : \forall S1 \exists I1 \exists tc \exists EV1 (S(S1) \wedge SI(I1) \wedge TCO(tc) \wedge AS(S1, I1) \wedge EE(EV1) \wedge CTD_{SE-EE}(I1, EV1) \wedge HCO\big(EV1, tc_{Flag(1)}\big) \rightarrow (FTCO(tc) \vee ITCO(tc) \vee RTCO(tc)))$$

(d) ***Visibility Constraints***—*Visibility Constraints* (*VCO*) put restrictions on service visibility during service discovery process. These constraints can be considered as the prerequirements for a service to be visible. Once the consumer binds with the provider, those constraints are of no use [1]. *Visibility Constraints* are of three types—*Awareness*, *Reachability* and *Willingness*. *Awareness* gives the information about the existence of provider and consumer, *Willingness* tells whether the parties are intended to interact with each other. *Reachability* is the condition when parties are able to interact with each other through message passing. All these conditions are disjoint to each other. When these preconditions for service visibility are satisfied, the *Visibility_Flag* is set. A formal definition of *Visibility Constraints* is

$$F9 : \forall S1 \exists I1 \exists vc \exists EV1 (S(S1) \wedge SI(I1) \wedge VCO(vc) \wedge AS(S1, I1) \wedge EE(EV1) \wedge CTD_{SE-EE}(I1, EV1) \wedge HCO\big(EV1, vc_{Flag(0)}\big) \rightarrow (Awareness(vc) \vee Reachability(vc) \vee Willingness(vc)))$$

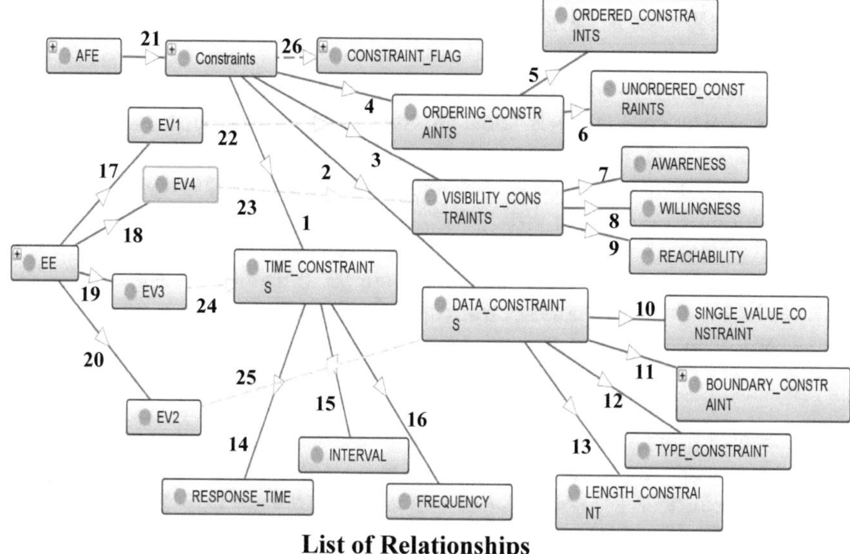


**List of Relationships**

**1 - 21. Has Subclass Relationship** **22 -25. hasConstraint Relationship**
**26. Has Flag Relationship**

**Fig. 1** Ontology graph for proposed constraints obtained from the OntoGraf plug-in of the protégé

## *4.2 Protégé-Based Implementation of Proposed Constraint Specification*

In this section, the proposed ontology-based constraints are implemented using an ontology editorial tool Protégé. A partial ontology graph for constraints given in Fig. 1 is obtained from the OntoGraf plug-in of the protégé. Here, the solid lines represent the classes and subclasses and the dotted lines represent the relationships between these classes. The figure has displayed that class ‘Constraints’ are subsumed by four subclasses—‘Ordering Constraints’, ‘Visibility Constraints’, ‘Data Constraints’ and ‘Time Constraints’. In addition, four subclasses of *Event Element* (*EE*)—‘EV1’, ‘EV2’, ‘EV3’ and ‘EV4’ are specified with four types of constraints using *hasConstraint Relationship*. Four types of ‘Constraints’ are further subsumed by respective subclasses as specified in Sect. 4.1.

## 5 Illustration of the Proposed Approach Using a Case Study

In this section, the proposed model is initially validated based on a case study specified in [13]. The case study is about mortgage application. This application has described services for a business process granting a mortgage. A client can apply for a mortgage at a branch office. This service entails the collaborative participation of people from the staff of three different organizations: a Branch Office, a Valuation Office and a Notary Office. The functionality of the service starts with two activities, (i.e. collect applicant data and value), which are performed by the Head of Risk of the Branch Office and an Appraiser from the valuation office, respectively, and which can be carried out in parallel. The cooperative task Mortgage Granting ends with the signing of the Title Deed by the notary, the Bank Manager of the branch, and the client.

According to this case study, the application is consisting of five services—'Mortgage Application Service', 'Component Valuation Service', 'Studies of Feasibility Service', 'Signing Deed Service', and 'Client Data Service'. 'Mortgage Application Service' has interacted with two services—'Studies of Feasibility Service' and 'Signing Deed Service'. Further, 'Studies of Feasibility Service' has interacted with two services—'Component Valuation Service' and 'Client Data Service'. These two services are invoked by 'Studies of Feasibility Service' in the same time. 'Component Valuation Service' has the capability to collect appraisal value. This functionality has performed by an appraiser of Valuation Office. Further, client data (payroll, unpaid and account data) are collected by invoking 'Client Data Service'. The functionality of 'Client Data Service' is performed by Head of Risk of Branch Office. The capability of 'Sign Deed Service' is to check the status of debt report and based on the status of debt report Head of Risk has created deed and that deed is signed by notary, bank manager and client. Table 3 has summarized the related services, consumer, provider, activities, data etc., of the case study.

**Table 3** Summarization of the mortgage application case study

| Facets of the mortgage application | Equivalent concept in the proposed model |
|---|---|
| Mortgage Application Service, Studies of Feasibility Service, Component Valuation Service, Signing Deed Service, and Client Data Service | Service (*S*) |
| Interface of all aforementioned services | Service Interface (*SI*) |
| Valuation Office | Actor; Provide Component Valuation Service |
| Appraiser | Actor; Perform Component Valuation Service |
| Branch Office | Actor; Provide Mortgage Application Service, Studies of Feasibility Service, Signing Deed Service and Client Data Service; and Consume Component Valuation Service |

(continued)

**Table 3** (continued)

| Facets of the mortgage application | Equivalent concept in the proposed model |
|---|---|
| Head of Risk | Actor; Perform Studies of Feasibility Service and Client Data Service |
| Bank Manager | Actor; Perform Mortgage Application and Signing Deed Service |
| Notary Office | Actor; Provide Signing Deed Service; and Consumer Studies of Feasibility Service |
| Notary Officer | Actor; Perform Signing Deed Service |
| Mortgage Client | Actor; Consume Signing Deed Service |
| Feasibility Study, Granting a Mortgage, Taken Decision, Agree and Sign Deed | Composite Action (*CACN*) |
| Collect Applicant Data, Collect Applicant Value, Feasibility study of client data and client appraisal value etc., | Simple Action (*SACN*) |

On triggering the event *Check_Feasibility_And_Status* at time *t1*, the 'Studies of Feasibility Service' composes with the 'Mortgage Application Service' at time *t2*. In this case, event *Check_Feasibility_And_Status* is specified with *Ordered Constraint*. Thus, there is *is_next* relationship between 'Mortgage Application Service' and 'Studies of Feasibility Service'. Further, 'Client Data Service' and 'Component Valuation Service' are composed with 'Studies of Feasibility Service' at time *t3*, when the event *Collect_Information* is triggered at time *t3*. In this case, event *Collect_Information* is specified with *Unordered Constraint*. Thus, 'Component Valuation Service' and 'Client Data Service' are composed with 'Studies of Feasibility Service' in parallel. Besides this, the 'Signing Deed Service' composes with 'Mortgage Application Service' by *follows* relationship (*Ordered Constraint*) at *t4*, after the event *Check_Feasibility_And_Status* is triggered at time *t1* on both 'Mortgage Application Service' and 'Signing Deed Service'. Figure 2 demonstrates this case.

## 6 Conclusion

This paper is aimed at specifying constraints related to SOA with well-formed syntax and semantics. The motivation behind the work is that effective occurrences of distinct observable facts related to SOA like service discovery, service execution and service composition depend on rigorous and precise specifications of related constraints. Yet, the existing-related research works focus on well-defined syntax primarily. Emphasis on expressing semantics are almost absent in these. The novelty of the proposed work is the representation of constraints with precise syntax and semantics in three important phenomenon of SOA—service discovery, service execution and service composition as well as data in services. Further, the proposed constraints are represented in meta-model level and are in middle-level ontology.

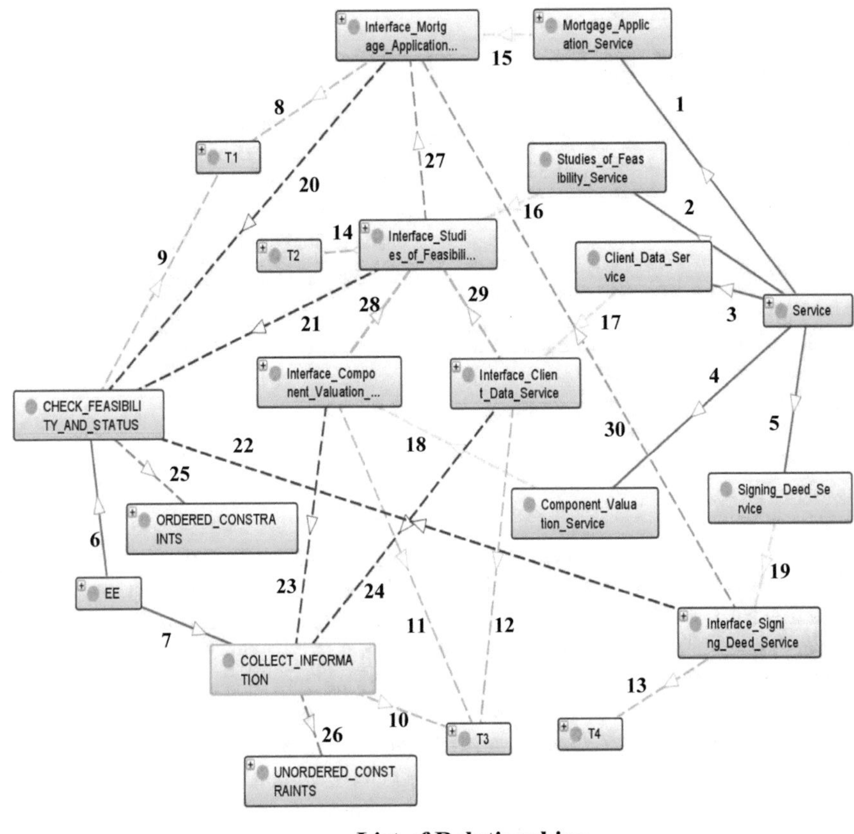


**List of Relationships**

**1 – 7. Has Subclass** **8 -14. Has Time** **15 – 19. Has Interface**

**20 – 24. Has Containment Event** **25 – 26. Has Constraint**

**27. is_next_to** **28. – 29. Unordered Relationships** **30. Follows**

**Fig. 2** Partial Illustration of the case study including different services and relationships between them based on distinct ordering constraints

Therefore, the proposed specification is capable to be reused in large numbers of applications in software service domain. Besides this, both user specified and system generated constraints are addressed through the proposed conceptualization.

Future work will include automation and validation of the proposed constraint specification. Extension of the proposed specification in other paradigms, such as agent oriented, component based etc., will also be a significant future work.

## References

1. Reference Architecture Foundation for Service Oriented Architecture Ver. 1.0, OASIS, Dec 2012. https://docs.oasis-open.org/soarm/soa-ra/v1.0/soa-ra-cd-02.pdf. Accessed 20 June 2017
2. Wang, Q., Li, M., Meng, N., Liu, Y., Mei, H.: A pattern based constraint description for web services. In: 7th International Conference on Quality Software (QSIC '07), pp. 60–69. IEEE, Portland, OR, USA (2007)
3. Banerjee, S., Sarkar, A.: Ontology driven approach towards domain specific system design. Int. J. Metadata Semant. Ontol. **11**(1), 39–60 (2016)
4. Banerjee, S., Bajpai, S., Sarkar, A., Goto, T., Debnath, N.C.: Ontology driven meta-modelling of service oriented architecture. In: International Conference on Communication, Management and Information Technology ICCMIT'17, University of Warsaw, Warsaw, Poland, 3–5 Apr 2017
5. Ngan, L.D., Jie, L.Y. Kanagasabai, R.: Dynamic discovery of complex constraint-based semantic web services. In: Fifth International Conference on Semantic Computing, (ICSC 2011), pp. 51–58. IEEE, Palo Alto, CA, USA (2011)
6. Guarino, N., Oberle, D., Staab, S.: What is an ontology? In: Staab, S., Studer, R. (eds.) Handbook on Ontologies, 2nd edn, pp. 1–17. Springer, Berlin, Heidelberg, Germany (2009)
7. Shen, J., Beydoun, G., Low, G., Wang, L.: Aligning ontology-based development with service oriented systems. Inf. Syst. **54**(2015), 263–288 (2015)
8. Horridge, M.: A Practical Guide to Building OWL Ontologies Using Protégé 4 and COODETools, Edition 1.3., The university of Manchester, 2011, https://mariaiulianadascalu.files.wordpress.com/2014/02/owl-cs-manchester-ac-uk_-eowltutorialp4_v1_3.pdf. Accessed 15 June 2017
9. Degwekar, S., Su, S.Y.W., Lam, H.: Constraint specification and processing in web services publication and discovery. In: IEEE International Conference on Web Services (ICWS, 2004), pp. 210–217. IEEE, San Diego, CA, USA (2004)
10. Mayer, W., Thiagarajan, R., Stumptner, M.: Service composition as generative constraint satisfaction. In: Proceedings of the 2009 IEEE International Conference on Web Services (ICWS '09), pp. 888–895, IEEE, Los Angeles, CA, USA (2009)
11. Croitoru, M., Compatangelo, E.: Ontology constraint satisfaction problems using conceptual graphs. In: Bramer M., Coenen F., Tuson A. (eds.) Research and Development in Intelligent Systems XXIII, The Twenty-Sixth SGAI International Conference on Innovative Techniques and Applications of Artificial Intelligence (AI 2006), pp. 231–244. Springer, London (2007)
12. Liu, S., Correa, M., Kochut, K.J.: An Ontology-aided process constraint modeling framework for workflow systems. In: 5th International Conference on Information, Process, and Knowledge Management (EKNOW 2013), pp. 178–183. International Academy, Research and Industry Association (IARIA), France (2013)
13. Noguera, M., Hurtado, M.V., Rodríguez, M.L., Chung, L., Garrido, J.L.: Ontology-driven analysis of UML-based collaborative processes using OWL-DL and CPN. Sci. Comput. Program. **75**(8), 726–760 (2010)

# Software Regression and Migration Assistance Using Dynamic Instrumentation

**Nachiketa Chatterjee, Amlan Chakrabarti and Partha Pratim Das**

**Abstract** Companies and organizations use the legacy software for decades to serve various purposes. During this journey, the software system travels through several change requests and amendments of functionalities due to the changing nature of business and other requirements. As a result, different methodologies and implementations employed over the time are often not at all documented. So, modifying or migrating those software systems become difficult due to lack of technical knowledge about their behavior. This difficulty is even more when there is no Subject-Matter Expert (SME). Here, we propose a technique to verify the unchanged functionalities of untouched modules of the modified application by comparing with the older version of the application. Sometimes, the number of functional behaviors become irrelevant as they are no longer required by the business. However, significantly large portions of legacy applications continue executing, untouched by any modification or customization, to serve tiny yet critical purposes. Stakeholders also remain reluctant to cleanup or migrate because only for finding out the active part or functionals scope of the application is very tedious and consumes lot of effort due to lack of knowledge or documentation. Here, we have devised a mechanism to assist the migration specialists to identify the active part of an application, associated files, and data used by the active code that help in building the new one with similar functionalities. We can also assist the performance engineer by detecting the resource leakage in the application.

N. Chatterjee (✉) · A. Chakrabarti
A. K. Choudhury School of Information Technology, University of Calcutta, Kolkata 700098, India
e-mail: nachiketa.chatterjee@gmail.com

A. Chakrabarti
e-mail: amlanc@ieee.org

P. P. Das
Department of Computer Science and Engineering, Indian Institute of Technology Kharagpur, Kharagpur, India
e-mail: ppd@cse.iitkgp.ernet.in



R. Chaki et al. (eds.), *Advanced Computing and Systems for Security*,
Advances in Intelligent Systems and Computing 897,
https://doi.org/10.1007/978-981-13-3250-0_12

**Keywords** Software migration · Regression · Dynamic instrumentation

## 1 Introduction

Many companies classically use large legacy systems for a long time, maybe for decades. Sometimes, these systems are stable enough and serve certain business-critical needs. Over the time, due to various change in policy and business strategy, several modifications can be employed on the application by various vendors. Due to this, the legacy code travels through a variety of software development methodologies adopted by different stakeholders, but sometimes legacy code evolved even without any software governance. Business models can also become obsolete as there may be a change in trend and strategy of the company, which may require further addition of new functional modules. Based on the ad hoc requirements from the business users software, developers implement the new features without any requirement or design documentation. It becomes even pathetic when this journey involves several different sets of developers, business users, and stakeholders. Change of company policy, business rules, and different requirement of business users continuously add new features to the system. To have an early deployment and to avoid unknown bugs, a developer always tries to incorporate the new features without touching the existing functionality. Due to this, the application size grows enormously without any clue of maintainability. Sometimes, the organization does not think about touching the application or doing any further modification on it.

But when the alteration required is a must for the existing functionality, the challenge is to ensure the quality in regression where the testing documentation and data are not available. In some cases, when developers rely only on code and business, the user can only certify if the visible output is as expected. Consequently, any issue reported in the modified application becomes tedious to debug because not enough knowledge is available about the code and data that are responsible for the observed behaviour. This can lead to difficulties for the regression tester and developer to avail the tool assistance to cross-check the functionalities in different checkpoints like comparing the function output against golden data, sequence of events, etc.

Sometimes, these new modules cannot be built upon the legacy system due to their requirement of sophisticated infrastructure and framework. As a result, application migration is an absolute need for the company. But as part of that, business-critical application module of the legacy system also needs to be migrated in the newer system. In this scenario, existing documentation and knowledge of the SME is the only key for the software vendor to identify the business rules and business-critical data, and then the ETL process implemented to plug-in it into the newer application. If the SME and documentation are not available for analyzing, a mammoth system is used to find out the active business rule is costly enough and not foolproof. Sometimes, it becomes costlier than building the newer application. Here, the migration specialist asks for some tools to help them focus on the actual area of the mammoth application.

So while working with the legacy application, the specialists ask for the tool support to enhance productivity during regression testing after modification, migration of application and support to enhance performance. Here, in this paper, we focused on the tool assistance to the legacy application specialists. We have discussed the existing proposals in Sect. 2 and then, we identified the target support that we want to achieve in Sect. 3. We described our solution approach and the implementation strategy in Sect. 4. Section 5 describes the observation and verification of the implementation and then, we concluded in Sect. 6.

## 2 Related Work

Several regression testing tools employed over the time for different purposes. Most of them evaluate the quality of software product in terms of their functional behavior.

Hwang et al. developed Java Dynamic Testing tool by using Instrumentation [1] to quantify productivity and other measurements of the software product. In this tool, they used the source instrumentation techniques to inject the additional code in Java source code at every iteration and checks. They used the source parser to detect the point of injection and based on certain rules, they injected the code to get quantitative data. After the instrumentation tool compiles, the source code executes to retrieve the data for computer manager to estimate the project effort as well as to calculate the productivity.

In demand-driven structural testing, [2] Misurda et al. proposed a demand-driven approach and established a framework to test programs based on execution paths to evaluate test coverage. Here, they defined Jazz, the prototype that test framework with the GUI, test planner, dynamic internment, and test analyzer. They also incorporated those features into the Eclipse Integrated Development Environment (IDE) [3].

Sometimes, testers want to validate the application behavior on performing a specific action while a module is in a particular state. Upadya [4] realized that while testing state-based systems, this is not usually possible with having classical quality assurance techniques as sometimes, the state transitions are either not externally visible/observable or, in some cases they are visible but not controllable, in real time. Dynamic instrumentation technique addresses this challenge, where the binary system under test is dynamically instrumented in such a way where transitions through various states of tests become observable and controllable. It is required to test such systems deterministically covering all different states and its transitions in an efficient and effective way. In this approach, they used dynamic instrumentation to detect the internal events of different states and state transitions that lends itself well for generating data driven tests and the ways that can be implemented for augmenting the test automation already in place.

There are also a couple of tools and strategies available to address the migration work. Various popular and efficient ETL tools facilitate the developer to setup the extract, transform and load data. But to start with the first phase extract, the developer needs to be confident enough about what data they wish to extract. For legacy systems, under consideration, are well documented or guided by experienced SME, and this task is quite easy for the business analyst to document the requirements supplied by the SME and forward it to the ETL developer. But when there are neither SME nor documentation, only relying on the code to derive the requirement from nature of the legacy system is difficult. Few approaches attempted to address this challenge to assist ETL developer to understand the extraction logic from the idea of critical business rules.

Business rules that organizations generally follow to perform daily activities evolve over the time and as a result, software aligned to that behavior also gets modified. Extracting or understanding these embedded business rules are very difficult when the encompassing software becomes large and aged. Furthermore, often business organization rely on the code base than any other documentation as because encompassing software changed multiple times without changing the corresponding documents. However, it is very expensive exercise to develop a generic tool to extract the business rules. Huang et al. [5] proposed a tailored solution to address this problem of extracting business rules, which includes variable classifications, program slicing, and hierarchical abstraction among other maintenance techniques. The proposed approach was developed and successfully experimented with couple of industrial applications.

Streitel et al. [6] also found that software systems contain dead code is an unnecessary burden for maintenance and it is often difficult to detect which parts of the software system can be removed. So, they attempted a semiautomatic, iterative, and language-independent approach to detect classes no longer used in any large object-oriented systems. This approach includes a very tedious iterative procedure aided by some tool support that executes on the class level to detect potential classes that can be entirely removed from the application. Further, they adopted the reflection mechanism of class loading to identify a missing link within the dependency graph between class that is being loaded and the class that performs the loading. It includes both static and runtime information about an application and aids developers detecting the code no longer used in their system. The list of classes then was prepared during the execution that contains at least one function that was invoked during execution. Generating this set can be easily achieved with a profiler. Some methods such as ephemeral profiling even allow production systems profiling. The developer has to manually search the source code for the mechanism that loaded at least one such class. Using this approach, the developer can definitely save some effort but cannot have the flexibility to rely on the tool without any manual intervention. Also, this approach can assist in the object-oriented application having reflection support. But unfortunately, large number of legacy does not belong to that family.

Tool Oovcde [7] uses CLang to parse C++ files. It finds all methods and functions, and determines with a call graph which functions are not called by another function. It outputs a table of the number of times each method or function is called. The table is viewable in a browser and its XML form also generated. These tools have certain limitations but it is still attempted well to detect the static and runtime dead code detection. To identify the runtime dead code detection, the tool use the source code instrumentation mechanism. Where the source code is instrumented with a special function to extract the execution counts and later, the tool analyzes the data to identify the dead code. So to use this tool, the application should have the ready source code with build infrastructure. Also, the application need to be in C++ only. Again most of the legacy application is not object-oriented, even sometimes, they suffer from the unavailability of source code or build infrastructure.

Gupta et al. [8] presented a path profile-guided partial unused code removal mechanism. This enables sinking for the elimination of deadness in frequently executed paths by augmenting additional set of instructions in infrequently executed paths. Their proposal of optimization is specifically suitable for special architectures because it varies upon generation of fast schedules in frequently executed paths by minimizing their critical path lengths. They also established cost–benefit, data flow analysis which uses path profiling report to establish the profitability of using predication driven sinking. By detecting paths along which an additional statement is introduced helps to determine the cost of a statement past merge point. Due to this additional dead code elimination is achieved by predication. The results of this analysis are incorporated in a code sinking framework in which predication-enabled sinking is allowed past merge points only if its benefit is determined to be greater than the cost.

## 3 Problem Statement

The available approaches address different purposes, but we do not see a clear addressing toward the ask of legacy application specialist for regression, migration, or performance tuning of their application. Here, we like to propose a tool to address the need for legacy application during regression and migration. The proposed tool will help with the below features:

- Retrieve application behavior
  - Golden Data—Functional Behavior
  - Sequence—Functional flow, Basic Blocks, Events flow
- Active module/fragment of code
  - Active image—Modules currently in use
  - Active function—Functions of Active modules
  - Active basic blocks—Identify active conditions
- Accessed data point by the code
  - File access
  - XML data access
  - Legacy database connection and transaction.

## 4 Solution Approach

Here, we are talking about the application suffering from lack of documentation or SME support, so it is quite clear that we need to identify the facts from application itself only. To discover these facts, we need to insert certain queries or concern in the application in such a way that the application may answer them during execution. This can be achieved by using code instrumentation and that can be done at three stages:

- Source code instrumentation where code is added before the program is compiled, using source-to-source transformation. Meta-programming frameworks like Proteus can also be used to insert extra code.
- Binary code instrumentation adds instrumentation code by modifying or rewriting the compiled code, through specially designed tools, either statically or dynamically.
- Bytecode performs tracing within the compiled code. It can be static as well as dynamic.

For our purpose, we will use dynamic/binary code instrumentation, where we do not need to touch the existing code. Touching source code is vulnerable and compilation infrastructures sometime are very tedious. Dynamic Instrumentation gives us the flexibility to include new analysis codes at any point of time according to the need of designer without changing the source code. Here we use PIN [9], a dynamic instrumentation framework, to provide an assistance to the programmer. We use dynamic instrumentation framework to analyze the legacy application by simply injecting the analysis routines in different levels, such as image level, function level and block level. From the call back of analysis routines we will capture the footprint of the execution.

### *4.1 Regression Assistance*

We assume here the application to be enhanced, need to be executed with all critical functionalities with the supervision of business users. From its execution, we will record the data points for each functional level, events like file read/write and it's sequence. The extracted record then persists in a file. We will then filter out the redundant data for the function level by de-duplication. These filtered data will be considered as golden data and may be supplied to the test framework to validate the new build.

In Fig. 1, we have defined the instrumentation strategy of our PIN tool for regression assistance. We used a map to hold the function name associated with a list of the arguments and return value set. Each function is instrumented to record its execution footprint. Before and after every function call, the instrumented method populate the set with argument list and return value. At the end of execution, we will get the

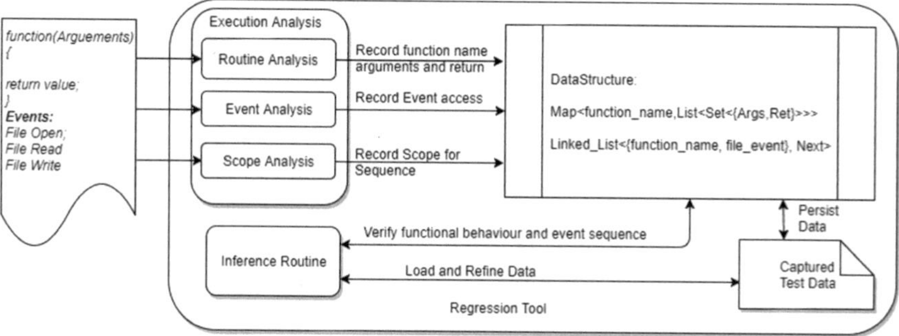


**Fig. 1** Regression tool strategy

complete map having all function name as a key associated with the list of argument and return value set for each call. The inference routine then uses the set compare to eliminate the duplicate set from the list. We use the set as a golden data to verify the execution of new build.

We assume here that the new build will have the same sequence of execution compared to the base application. If the execution sequence differs, then this tool will not be able to completely address the scenario, whereas this will still retrieve the golden data. Our other implementation [10] can address this to do the unit testing with these golden data in different functional level. That tool can be used for testing desired function with golden data by use of simple XML configuration. This tool can also be used for newly introduced function to test against the golden data that we have discovered.

Upadya [4] used event-driven testing approach using dynamic instrumentation to test the sequence of the file events. Here also we used a similar mechanism to identify the events and its access sequence from the function. We captured the resource access events and the scope to identify the sequence. We have defined a linked list to capture the sequence of method execution and event access. The instrumentation function for retrieving golden data also populate this access list and record the footprint during function entry. Similarly, the file open, read, and write events instrumented in such a way that the callbacks can populate the linked list with the event access footprint. Finally, we get the sequence of functional execution and the events. We can verify the new build against this sequence. Using this tool, we may also detect the difference in sequence and assist the developer to identify if that discrepancy is intended or not.

## 4.2 *Migration Assistance*

We have studied the tools and approaches for migration assistance and analyzed the above-mentioned problems of migration phase for various legacy applications. To address the main tasks like generating the active code, active data, and their relationship, they fall short. So here, we concentrate on the implementation of our tool to assist the developer to focus on the active control flow of legacy application to identify the active code and resource access.

**Active Code Detection** can be done by analyzing the execution sequence that we have identified already as part of regression. However, here, we are more interested to understand the functional details of the application in terms of functions belonging to images or library and the associated data. After executing all known functionality, we will be able to discover all active code in granular level responsible for those functionalities. Now these information can also be maintained in functionality level. The migration developer can use these information to easily filter out the active code from the mammoth application. Even they can review and rewrite the existing active code responsible for a particular functionality. Also, this tool can assist developer to maintain clear modularity between different functionalities.

Here, we defined the migration tool design in Fig. 2, where we have used a linked list similar to the regression tool to capture the function sequence only. The linked list is further associated with the image in a map. Here, we are not capturing the argument and return value of the functions. Each element of function list is a structure where function name is captured along with a list of resource access from a function. Each function is instrumented to generate the scope sequence and associated with the list of resource access. Every resource access event captured and stored in this structure along with the function from where the access occurred.

**Active Data Detection** can be done by recording the data accessed from those active codes. During the execution of the application, we have instrumented the data

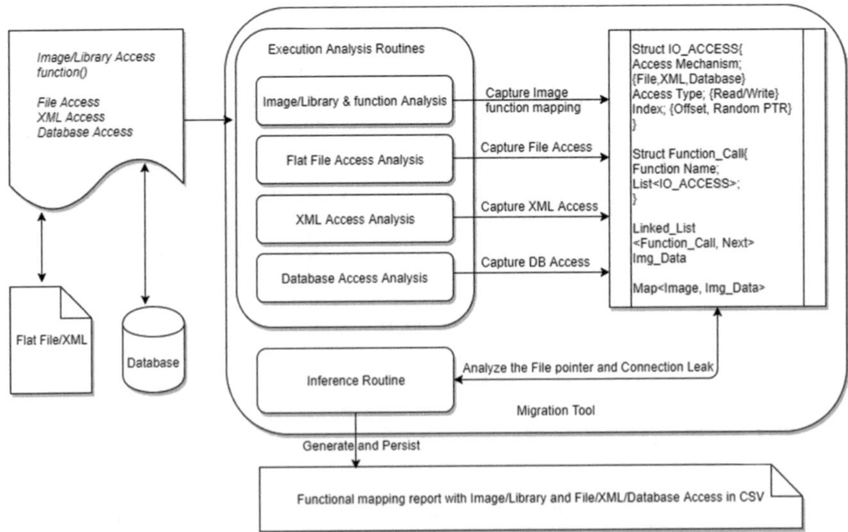


**Fig. 2** Migration tool strategy

access events like file read or write, database connection and data read or write. From these instrumentation callback we have acquired the information about what data accessed from the active code. These information help developer to identify the active data associated with the application. To capture resource access we developed a data structure to record the access mechanism (flat file, XML file, or database), access type (Read, Write, Open, and Close), and the index or offset. Here, we have instrumented the flat file access functions to identify the access events. So with this, we can now populate the data structure to discover which file accessed from which part of the application along with the access type and associated image. Now this has been further enhanced with the XML node access, where we have instrumented the XML node access function and the callback also populate the same data structure with an appropriate mechanism and access type. Similarly for the other XML functions like XML file open, XML element read/write, etc., we have captured the information in our data structure. We have also instrumented the database access functions to identify the live data. Here we instrumented the init, fetch, query, close, etc. The callback of those instrumentations also populates the resource access structure.

### 4.3 *Performance Assistance*

With the above-extracted data from the migration tool, we analyzed the sequence of execution of different file and data access. We have noticed that in few scenarios, the connection not closed in the application causes connection leak. So here, we have introduced the scope variable during the execution analysis. That scope has been tagged with the initialization operation with the input output resources. So whenever the resource opened for the access, we have captured and recorded it in a data structure and then deleted the record while closing the resource. Every time we exit a scope, we check the record against that scope. If the resource list against the scope is not empty, we detect a resource leak. We also can initiate a breakpoint to assist the developer to identify the exact location in the code. By eliminating resource leak, we can improve the performance.

```
//Data Structure to track scope
  map<scope_id, List<IO_Access>> active_access;
  scope_id_global_var;

//Instrument Analysis routine for each scope begin
  instrument_begin_scope(scope_id)
  {
    scope_id_global_var = scope_id;
  }

//Instrument Analysis routine for each resource initialization
```

```
  instrument_resource_init(IO_Access resource_id)
  {
    active_access[scope_id_global_var].push_back(resource_id);
  }

//Instrument Analysis routine for each resource close
  instrument_resource_close(IO_Access resource_id)
  {
    active_access[scope_id_global_var].delete(resource_id);
  }

//Instrument Analysis routine for each scope end
  instrument_begin_scope(scope_id)
  {
    if(active_access[scope_id_global_var] != empty list)
        //record leaked resource and initiate beakepoint
  }
```

## 5 Observation and Results

We have verified our regression tool implementation to identify golden data to use them as test data for the modified application. Our test suite contains various program with different locality of data as in Table 1. We are able to detect the primitive data and other serialize data pointers. We have also used various benchmark code to test this tool. We are able to detect the functions, arguments, return values, and their scope. We observed few functions use third control variable in it such as time stamp are not deterministic in nature, we identified them as unreliable data set.

We have also verified migration tool implementation on sample application having the suitable features like complex file access, XML access, and database access. We have prepared a test suite with the sample program having the features as in Table 2, that we wanted to test with our tool. We have also developed our test suite with tinyXML library to access the XML file and MySQL library for database access. Where an application opens the file of different type, parse the XML file and access different node, elements, and attributes. We have identified the events and successfully tag them with the functional implementation.

In terms of database, we have considered the mySQL integrated with the C program by using the mySQL library. We have successfully identified the database access events like initialization, execute query, fetch number of rows, and results. Also, we can detect the close connection and result set. When the connection or result set not closed within the scope, we can detect that as a leak. We have also identified the unclosed file by use of our performance tool but in the case of tinyXML,

**Table 1** Test suite to detect golden data for regression tool

| Function | Arguments | Return | Scope |
|---|---|---|---|
| Macro | × | × | × |
| Global Function | ✓ | ✓ | NA |
| Static Function | ✓ | ✓ | NA |
| Constructor | ✓ | NA | NA |
| Overloaded Constructor | ✓ (part) | NA | NA |
| Copy Constructor | ✓ | NA | NA |
| Destructor | × | NA | NA |
| Inline Function | ✓ | ✓ | ✓ |
| Friend Function | ✓ | ✓ | ✓ |
| Member Function | ✓ | ✓ | ✓ |
| Overloaded Member Function | ✓ | ✓ | ✓ (part) |
| Virtual Member Function | ✓ | ✓ | ✓ |
| Overridden Member Function | ✓ | ✓ | ✓ (part) |

✓—Success, ×—Failure, NA—Not Applicable

**Table 2** Test suite for active data and active code

| Test scenario | Expected result | Status |
|---|---|---|
| Program involving multiple library | Detect active libraries | Success |
| Multiple header and functions | Detect active functions | Function detected<br>Header not detected |
| Program access files | Detect File Open<br>Read/Write/Index<br>File close<br>File error<br>File not closed | Success<br>Success<br>Failed<br>Success<br>Success |
| Program access XML using tinyXML | Detect XML Open<br>element/node/attribute<br>File not closed | Success<br>Success<br>Success |
| Program access Database using MySQL | Detect DB Connection Open<br>Execute query<br>Fetch result/row count<br>Connection error<br>Close Connection/Results<br>Connection/Results Leak | Success<br>Success<br>Success with some data types<br>Success<br>Success<br>Success |

we observed that the file managed by the loader and hence closing operation is not explicitly available.

We have also chosen benchmark code to test the different types of file access and is able to successfully detect the operations.

## 6 Conclusions

In this paper, we have derived a tool support to address various needs during regression test of application enhancement, migration of application, and performance tuning.

Our regression tool is able to generate the unique set of unit test data for function level and the data set may be used to verify the new build. We have also generated the sequence of other resource access events to verify the behavior of enhanced application. We have identified few nondeterministic functions in the application where output differs from one run to another for the same input set. We defined them as unreliable data set. In this case, we want to further enhance our tool to identify the data flow in basic block level to verify the execution pattern.

We discovered the active code flow and the data accessed from it using our migration tool. The implementation clearly identified all type of file access. We have successfully identified the tinyXML and mySQL access. However, the other kinds of XML and database support is restricted. However, the library support can be further enhanced with the configuration modification of the tool.

We have detected the resource leak in the application and also identified the scope of exact root cause. This will assist the performance engineer to fix it and improve the code quality.

Our future work would be to increase the library support of the migration tool and extend the tool to integrate with GUI support along with other instrumentation tools.

## References

1. Hwang, S.-M., Lee, J.: Design and implementation of Java dynamic testing tool using instrumentation. Indian J. Sci. Technol. **8**(1), 475480 (2015)
2. Misurda, J., Clause, J.A., Reed, J.L., Childers, B.R., Soffa, M.L.: Demand-driven structural testing with dynamic instrumentation. In: ICSE'05, St. Louis, Missouri, USA (2005)
3. Eclipse integrated development environment. http://www.eclipse.org
4. Upadya, V.: State-based testing using dynamic instrumentation a case study. In: PNSQC 2012 Proceedings, pp. 1–6 (2012)
5. Huang, H.: Business rule extraction from legacy code. In: COMPSAC '96 Proceedings of the 20th Conference on Computer Software and Applications, p. 162 (1996)
6. Streitel, F., Steidl, D., Jrgens, E.: Dead Code Detection on Class Level. CQSE GmbH, Garching bei München, Germany (2014)
7. Detecting dead C++ code. http://oovcde.sourceforge.net/articles/deadcode.html
8. Gupta, R., Berson, D.A., Fang, J.Z.: Path profile guided partial dead code elimination using predication. In: PACT '97 Proceedings of the 1997 International Conference on Parallel Architectures and Compilation Techniques, p. 102 (1997)
9. PIN tool user guide. https://software.intel.com/sites/landingpage/pintool/docs/81205/pin/html/
10. Chatterjee, N., Bose, S., Das, P.P.: Dynamic weaving of aspects in C/C++ using PIN. In: HP3C-2017 Proceedings of the International Conference on High Performance Compilation, Computing and Communications, pp. 55–59 (2017)

# Author Index

**A**
Arun, Priti, 61

**B**
Bajpai, Shruti, 143
Banerjee, Shreya, 143
Bhattacharjee, Shuvadeep, 117
Biswas, Tanmay, 117

**C**
Chaki, Nabendu, 105
Chaki, Rikayan, 73
Chakrabarti, Amlan, 117, 159
Chatterjee, Moumita, 3
Chatterjee, Nachiketa, 159
Chavan, Bir Singh, 49

**D**
Das, Partha Pratim, 159
Das, Rajib K., 131
De Kumar, Bikram, 73

**G**
Ghosh, Sumonta, 15
Gupta, Savita, 49

**K**
Kaur, Damanjeet, 61
Kaur, Simranjit, 61
Khatua, Sunirmal, 105, 131

**M**
Mandal, Sudhindu Bikash, 117
Mondal, Sudip, 105
Mukherjee, Aradhita, 105

**P**
Pal, Anita, 15
Ptošek, Vít, 91

**S**
Saeed, Khalid, 27
Saha, Debasri, 117
Saha, Sanjoy Kumar, 73
Sarkar, Anirban, 143
Sen, Biplab Kanti, 131
Setua, S. K., 3
Singh, Jaskirat, 49
Singh, Sukhwinder, 49, 61
Sinha, Aniruddha, 73
Slaninová, Kateřina, 91
Szymkowski, Piotr, 27


R. Chaki et al. (eds.), *Advanced Computing and Systems for Security*,
Advances in Intelligent Systems and Computing 897,
https://doi.org/10.1007/978-981-13-3250-0

Printed in the United States
By Bookmasters